Praise for Timothy Close

My Coast Guard Option is a valuable read and welcomed perspective that allows one to walk in the shoes of a Coast Guard leader who has fought the good fight.

Tim Close's insider view speaks to the heart of leadership initiative, trust in training, working the problem, while always keeping the mission and people first.

Gary Weeden, Captain, U.S. Navy Chaplain Corps (retired) Former Chaplain of the Coast Guard

Tim Close didn't just choose the Coast Guard—he helped define it through a superlative career. From Academy jitters to SAR storms, his leadership blends calm, candor, and just enough irreverence to keep the mission and the crew afloat.

Joel Whitehead, Rear Admiral, U.S. Coast Guard (retired)

To my family, classmates, and fellow Coasties

MY COAST GUARD OPTION

MY OPPORTUNITY TO FIGHT THE GOOD FIGHT

TIMOTHY CLOSE

Captain, U.S. Coast Guard (RETIRED)

MY COAST GUARD OPTION

MY OPPORTUNITY TO FIGHT THE GOOD FIGHT

Published by Tactical 16 Publishing

Colorado Springs, Colorado

www.Tactical16.com

ISBN: 978-1-966413-11-0 (paperback)

ISBN: 978-1-966413-10-3 (hardcover)

ISBN: 978-1-966413-12-7 (Ebook EPUB)

Contents

Preface

I got a call at about 0130 on a Sunday morning. We had another SAR case. The command center relayed the key information: four men in a 22-foot boat who went out fishing Saturday morning were overdue. They were well overdue, and the wife of the boat owner (and one of the four missing men) was very concerned. She said they were never this overdue, and her husband always called as soon as the boat got back into port. She had no idea where they went fishing, other than it was offshore. At some point in the evening, she had contacted a friend of her husband who frequently went fishing with him and who was an experienced boater, but he did not know where they had gone. The command center also spoke with the friend, and he indicated that the GPS on the boat had numerous locations identified well offshore where they had success fishing on other trips, but he did not have that information stored anywhere else. And the command center passed the information that two of the four men were current NFL players. While that fact was not important relative to the SAR effort, we all knew that the case would attract media attention that other cases under similar circumstances would not.

I could tell immediately that the weather had deteriorated because I could hear the wind blowing hard, but the Command Center added the details. A sunny, breezy Saturday with the air temperature in the sixties and one to two feet waves had changed that evening to strong winds, fully overcast skies, temperatures in the forties, and ten to twelve feet waves offshore. We were aware from the start that the sea conditions and the weather made this case a fight against the clock.

Introduction

This is my memoir of my thirty years in the U.S. Coast Guard from 1982 to 2012. I wanted to capture some experiences and share some perspectives. But understand clearly, these are my perspectives: correct, incorrect, jaded, or otherwise. Hopefully, you will find them interesting and valuable. The times may have been different, and the technology certainly was, but the human emotions, desires, motivations, and stressors were mostly the same as they have always been and will always be.

Along the way, I had the privilege of making some close friends and working with some outstanding people. I learned a lot about leadership; some by witnessing what some excellent leaders did, some by watching and noting how not to handle things. I got to participate in some things I never dreamed of doing. I also got the chance to make a difference, and ultimately, that is what mattered the most.

When I was in third or fourth grade, we were given an assignment to create a seal of some sort that represented us individually, akin to the Great Seal of the United States. I don't remember what I drew specifically, but I remember struggling with a motto. I asked my dad for some help, and he sent me to retrieve the Bible from their bedroom. He opened it to Timothy I, pointed to the words, and suggested, "To fight the good fight, to have faith and a good conscience." It fit nicely in the drawing and fit perfectly for me. I remember being impressed and thinking at the time that Timothy from the Bible sure must have been a smart guy. It was only years later that I learned that those words were written <u>to</u> Timothy and not <u>by</u> Timothy. It probably actually fit better that way! But that motto stuck with me my entire life, and I embraced it fully.

During my career, I developed what I called the "Coast Guard option." It referred to a combination of buying into what the Coast Guard really was about, accomplishing the missions, the concept of using your "trained initiative," being motivated to not simply stand by and do nothing, and having a sense of doing what was right regardless of the consequences. It was about fighting the good fight. Sometimes, the "Coast Guard option" included exercising authority, whether it was authority granted to the Coast Guard in general in law and regulation or the legal authority granted to me based on the position I held at the time. Over my Coast Guard career, the "Coast Guard option" became increasingly significant as I was promoted and given positions of increased responsibility. But at the start, it was much simpler.

ONE

Just a Kid

I was born in 1960 in Massachusetts. My dad was on active duty with the U.S. Air Force, and we were stationed at Westover Air Force Base at the time. My parents were both from Pittsburgh, Pennsylvania, and had very different but challenging childhoods.

My father was born in 1932 and was an only child. His biological father abandoned him and my grandmother when he was still very young. His mother eventually filed for divorce and remarried a man named Bernard Richard Close. He became my dad's father and my grandfather in every way except biology. So complete was the transformation into this new family that my father's name was changed to Richard Bernard Close, completely obliterating his previous name and identity. He was adopted by his new father, although the official paperwork wasn't done until my father was in college and about ready to be commissioned as a Second Lieutenant in the U.S. Air Force. I suspect the lack of a paper trail prior to that was problematic for the Air Force, especially when it involved a complete name change, and it had to be formally fixed.

My grandfather was drafted into the Army during World War II despite being thirty-four years old, married with a child, and a steelworker at one of the huge steel mills that Pittsburgh was known for, and which were so critical to the war effort. My dad suspected that my grandfather pissed off someone in the union who then had a word with the Draft Board. In his pre-teen years during the war, my dad was certainly aware of what was happening and the stresses of having someone close to you engaged in the fighting in Europe.

My mom was born in 1934 to a family of Italian immigrants. Her mother, my Nana, was born in Italy and came to the U.S. as an infant. Her family settled in Pittsburgh and grew roots there. By the time my mother was born, she had two uncles, four aunts, and her Italian-speaking grandparents, all living near each other. My Nana, with my mom and her sister in tow, left her husband after one or more incidents when he had hit her. It was a different time back then, and divorce was frowned upon, especially for devout Catholics, but my Nana was a stubborn Italian and had no tolerance for being assaulted.

She moved back in with her parents and eventually met and married John Augustine, who also became my mother's father and my grandfather in every way except biologically. Another Italian immigrant, he was a hard-working carpenter and a good man. He took great pride in the buildings and structures he helped construct all over the Pittsburgh area. It was his sweat equity in the region. Possibly because of his limited education, he would frequently instruct me to "think deep thoughts." I tried to take his advice to heart as he was a simple man, but wise.

My parents met when they were both in college, introduced by mutual friends who were dating. It was a blind date. The funny part was that the date was at a "country" themed event that encouraged everyone to dress the part. My dad had to borrow a pair of blue jeans (he referred to them as "dungarees") from a friend. They were so big that he had to use a length of rope as a belt to keep them up. It may have been the only time in his life that he wore a pair of jeans, a matter of pride to him, but that blind date was the occasion that sparked a sixty-plus-year marriage.

My father graduated from the University of Pittsburgh on an ROTC scholarship and was commissioned a Second Lieutenant in the U.S. Air Force. In the several months between graduation and formal orders from the Air Force, he worked in the steel mill alongside my grandfather. As soon as my mother graduated from Mount Mercy College (now Carlow University), my parents were married. As they were driving to the Poconos for a short honeymoon, my mother said she was opening envelopes with money they received as wedding gifts, hopeful that they would have enough to pay for the honeymoon!

After a few duty stations, various training programs, and having a daughter, Lisa, my father was transferred to Westover Air Force Base in Massachusetts. I was born in Springfield, Massachusetts, in 1960.

When I was a year old, my dad got orders to Germany. By that point, he was working in photo intelligence. With the Cold War in full swing, Germany was where he was needed. It's hard to fully grasp the impact of orders like that in 1961. My parents left both of their families behind and headed to Germany with my older sister and me, knowing that they wouldn't likely see their parents for three years. Airfare was expensive, and no one had the money to fly back and forth. Phone calls were infrequent and challenging because of the limited long-distance phone line capacity and time zone difference. My younger sister was born while we were in Germany, which must have presented another set of challenges with no family support to assist.

After that, we were transferred back to Massachusetts, where my dad was assigned to help manage the Air Force ROTC program while simultaneously working on his master's degree and PhD in political science. Around that time, my older sister got sick with what turned out to be brain cancer. To deal with that problem, my dad turned down a promotion and orders to Japan and parted ways with the Air Force. We moved to Pittsburgh, where he started a teaching career at Waynesburg College, about an hour south of Pittsburgh. We lived with my mom's parents, and I started second grade there while we waited for our house to be built in

Waynesburg. Meanwhile, my older sister was seen and treated by a litany of doctors.

Waynesburg was a small town of fifty-three hundred people in a rural county in the southwest corner of Pennsylvania. It was uniquely a coal mining area and a small college town, and it became home quickly. We lived on a street with several other kids my age, and we spent a lot of time playing wiffle ball, baseball, basketball, and football. One of the neighbors with kids my age had played for the Steelers briefly and he would join our football games as the quarterback for both sides.

While all this was happening, my older sister's cancer was winning. She underwent several brain surgeries that didn't stop its spread but left her partly paralyzed as if she had had a stroke. She passed away too young in August 1969, at a time when non-surgical cancer treatment was only in its infancy. I saw the stress and anguish in my folks and only years later marveled at how they had clung to each other in such tough circumstances that statistically caused so many families to fracture. Their Catholic faith was strong and remained an example for me. I also remember how the community responded so warmly to this tragedy, especially for people who weren't from there and were still relatively new to town.

But life went on. It was school, sports, trombone practice, and friends. It was small town America where everyone knew almost everyone else, and if you didn't, it was guaranteed that you had several mutual acquaintances. You could walk just about anywhere, and it was safe if someone offered you a ride.

My friends and I spent a lot of time playing up on "the hill." We lived in the last house on the street and beyond our house was nothing but a rough trail leading up the hill and into the woods. We played Army all the time. We dug foxholes, which were a lot harder than we thought they would be. We pretended to shoot at Nazis and other enemies. We made parachutes for our GI Joes and set them up behind rocks for ambushes. When the weather was bad or it was extra hot outside, we set up our little plastic soldiers and staged mock battles inside. And then we would go outside and toss a base-ball or play wiffle ball. We had wiffle ball variations for as few as two players and up to as many friends as would join us. And that was interspersed with organized Little League games. It was a fun child-hood overall.

As a kid, I would occasionally pester my paternal grandfather, the steelworker, to tell me about World War II. He would usually just quietly change the conversation. But one day, I was especially pestering him and asked what his rank was. He hesitated, then said he had been a Sergeant but that they had busted him back to Private First Class (PFC). Not to be satisfied with that, I pressed him on why he had been busted. He hesitated again, then told me he hit someone. I pressed him more.

"Why'd you hit him, Grandpa?" He hesitated longer, and I could tell he was looking for the right words. Finally, he just said that the guy did something that got someone hurt. He added that a few days later, he was offered his stripes back, but he told me that he told the higher-ups to keep them. That story stuck with me. He had been thirty-six or thirty-seven when that incident happened and was likely the "old man" in his unit. Plus, he was a sergeant and had experi-ence. With that, I stopped pressing. It was obvious even to a kid that I had pressed him enough. He remained a PFC for the duration of the war, and that rank is on his grave marker. He clearly had no regret for hitting the guy but had recognized the Army's offer of reinstatement to Sergeant as their way of saying he had probably been right. But he stood fast. I suspect that he would have punched the guy again if he had to relive that moment. And that from a guy who was always calm and quiet, who never raised his voice, and who baked excellent cinnamon raisin bread.

Elementary school morphed into middle school and the years flew by. A few weeks after my eighth grade graduation (a big deal in Waynesburg because I suspect that was traditionally as much educa-tion as many people received), we moved across the state to a little town in the Lehigh Valley, Macungie. My dad had taken a position teaching at Kutztown State University, and so my parents, my younger sister, and I trekked across the state. New school, new friends, new neighborhood, and a bevy of classmates with Pennsylvania Dutch last names. No horses and buggies and no plain clothes; you had to drive the twenty-five minutes to the Kutztown area to see that, but lots of Pennsylvania Dutch ancestry and customs.

I showed up for the first day of school not knowing anybody, so I just hung back a bit and watched. It was ninth grade at a school that contained seventh, eighth, and ninth grades, and it was an odd "open concept" school, a new thing in the seventies that didn't work well then and mercifully went to the scrap heap of history. I focused on academics, my trombone playing, and freshman football, and soon enough, made some friends.

My sophomore year was at Emmaus High School (tenth through twelfth grades), and it combined kids from two feeder middle schools who all seemed to know each other. Again, I hung back and watched and focused on academics, my trombone playing in the jazz band, and JV football. I really liked playing football and quickly became the starting strong safety.

While the JV team played its own games, each week, two or three JV players were invited to suit up for the varsity game and maybe play on special teams. When my turn came, I found myself on the varsity kickoff team. Down the field I raced, and to my surprise, no one even tried to block me. I made it downfield quickly, dove at the ball carrier, got lucky, caught his foot, and tripped him up. Only when I got up did I realize that it was a tackle inside the twenty-yard line, a coveted achievement, albeit one that I felt was due largely to luck and a slippery dew-covered field. By the end of the season, that tackle earned me an invitation to do it again during the big annual varsity rivalry game against Whitehall High School held every Thanksgiving morning. Only this time, the other team's ball carrier was a big kid named Matt Millen, who went on to play for Penn State and several NFL teams, where he earned four Super Bowl rings. Oh, and he ran right over me on the opening kickoff and nearly scored a touchdown. You win some, and you lose some, but my passion for the game never faltered.

High school was fun, but felt very transient to me. I wasn't from there and had no intention of staying in the area. I knew I was going to college, but never once considered any of several decent schools in the area. I was looking elsewhere but had no idea what I wanted to do, let alone what I wanted for a major. My mom suggested I consider engineering because I was good at math and physics. I was a pretty good student overall, which did little to help narrow the choices of majors for me. I had a lot of different groups of friends with different interests. I could lift weights with other football players in the afternoon, then go to band practice in the evening, then go home and hit the books as easily as breathing.

Eventually, I felt a hint of recognition that the military might be a good fit for me. My grandfather had been a soldier. My father had been in the Air Force. To me, the Air Force sounded interesting, and the thought of being a pilot seemed intriguing. But the Army just sounded like a better fit deep down. So, I started the application process for West Point, the United States Military Academy. I filled out forms and submitted requests for an appointment to one of the Pennsylvania Senators. To hedge the bet, I applied to the other Pennsylvania Senator and asked for an appointment to the U.S. Air Force Academy. Never heard back from that guy, but got an interview for a West Point appointment.

Meanwhile, I applied directly to West Point to gain their approval. The football coaches seemed interested in me for reasons that I still don't understand. I was a decent high school football player, but not a star by any means. And I was only 5' 10" and 150 lbs. Maybe they felt that my grade point average would somehow boost the team's overall GPA. Maybe they thought I would grow a few more inches, beef up another thirty pounds, and become a stud (none of which happened). Even though I did not yet have the necessary political appointment, they assured me that they tracked which Congressional Representatives had and had not used all their appointments and would guarantee that I got one. As it turned out, my local Congressman had several appointments to use and graciously gave me one without the hassle of an interview or lengthy application process. By early December, I knew I could be headed to West Point.

On the evening of 14 December 1977, I got home late after an evening meeting or rehearsal at the high school, probably a Key Club meeting, which tended to be in the evenings. My mom said that an assistant football coach from the U.S. Coast Guard Academy had called. He wanted to know if I was planning on applying and would call back after I got home. I had a few minutes to dig through a big pile of college booklets, brochures, and pamphlets that had arrived unsolicited over the past few months. Sure enough, there was a booklet from the Coast Guard Academy in New London, Connecticut. I had never heard of it. But I flipped through the booklet, was mildly intrigued, and decided to hear him out when he called. He tried to sell me on the football team, a Division III school that played other small New England college teams. He explained how they started several freshmen the year before, and I could compete for a varsity position as a freshman.

While he was selling the football program, I was asking questions about academics. I learned that, ironically, he was a West Point graduate who had served his five-year commitment and left the Army to become a football coach. That spurred a lot of questions about West Point. By the end of the conversation, I told him that I would apply to the Coast Guard Academy. I had nothing to lose. He was glad to hear it and told me that I had to have the card in the back of the booklet filed out and postmarked by the following day, as that was the deadline!

I filled it out and my mom kindly drove it to the post office the next day. The response to that card was a larger application packet that arrived over the Christmas holidays. Lots of forms and essays about why I wanted to go to the Coast Guard Academy. It also asked which major I was interested in studying. The form pointed out that the selection of your major was not binding. I reviewed the limited majors (I believe they offered nine different majors at the time) and selected "Marine Engineering," which was related to the design of ships and their engineering plants. I had no real idea what that major was all about, but I felt it made me sound a bit more gung-ho for the Coast Guard, so I checked that box. In the process, I learned that the Coast Guard Academy did not use Congressional appointments and that all appointments were tendered directly from the academy. That was one less step.

I heard back from them in what seemed like two or three weeks. I had been accepted and had been offered an appointment. I was in at the Coast Guard Academy.

Now came the hard part: making a decision that felt like it would affect the rest of my life. I had applied to a few other civilian schools, too, but wasn't very serious about them. I was leaning toward engineering programs, but the basic engineering disciplines seemed uninteresting, with one exception: biomedical engineering. It was a relatively new field, but was offered at several colleges and universities.

One program that I applied for and may have jumped at if accepted was at Rensselaer Polytechnic Institute in Troy, New York. They offered a six-year program that combined a Bachelor of Science degree in biomedical engineering and medical school. It was a highly competitive program that went all year long, including summers, and students had to maintain a 3.0 grade point average. It sounded stressful and challenging but really interesting. I wasn't accepted. However, they did accept me into their regular biomedical engineering program. I was a little disappointed, but that made my choice easier.

It came down to West Point or the Coast Guard Academy. I knew that neither would be easy, but had no doubts I could handle either one despite a lot of butterflies in my stomach. West Point was more familiar, and I was inherently more familiar with the Army's mission and proud traditions. Part of me was still that kid playing GI Joe, digging foxholes and pretend shooting at Nazis. It fit. The Coast Guard Academy was different. It was smaller. Heck, the entire

Coast Guard was small. The majors were limited. The reputation was not widely known. I never knew or even heard of anyone who was in the Coast Guard.

But then I looked at the missions of each service. It was 1978, and the Cold War was still in full swing. The Vietnam War was over, and it seemed to me that the Army was focused on stopping the USSR should it attack Western Europe. It was the Fulda Gap, and tactical nukes, and missiles in silos. The Army's mission was all about training and practicing for what everyone prayed would never happen. With the limited information I had available, the Coast Guard appeared to be executing their numerous missions daily instead of just training for them. They were doing search and rescue. They were stopping drugs from getting into the country. They were providing port safety and manning lighthouses. They had icebreakers that were heading north and south on actual missions. Those sounded like good fights to me.

I had only visited the Coast Guard Academy once, and it was at the invitation of the head football coach. He seemed to be a great guy, and he arranged for a freshman cadet (also from Pennsylvania) to take me with him to a few classes and lunch. I found the people to be friendly, but the facility to be bleak. It may have been the dreary weather that day, but the image was not convincing in itself.

I had visited West Point three times, including a stay with one of the assistant football coaches, and was given tickets to a football game the following day. At least it was sunny there!

I felt a lot of pressure as if I was cutting the cord and totally moving forward on my own. That could not have been further from the truth, as my parents were exceptionally supportive, but it still felt like I was being asked to make an irrevocable decision about the rest of my life. That feeling was correct, but not in the way I thought as a young, naïve high school senior.

In various discussions and material about West Point and the Coast Guard Academy, two statistics jumped out and stuck with me. At that time, roughly 15% of West Point graduates stayed in the Army for a full twenty-year career. In contrast, roughly 85% of Coast Guard Academy graduates stayed in the Coast Guard for a full twenty-year career. Even as a kid, that contrast seemed significant. In simple terms, I felt there must have been something about the Coast Guard that they liked that kept them around. Maybe it was about the missions, the good fights.

I took the Coast Guard option. For the first time, but not the last, I took the Coast Guard option.

TWO

The Coast Guard Academy and the Class of 1982

The first few days at the academy were a whirlwind, probably exactly as they had always been for every person before me. But my head was spinning, and I had no idea which end was up for several days. It was called "Swab Summer," and all of us "swabs" were getting indoctrinated quickly. It was truly like drinking from a fire hose. Early on the second night, I got sick, probably just due to the initial stress and exhaustion. I spent the night in the clinic with a few other classmates. After a good night's sleep, I was better and rejoined my Swab Summer platoon. Wednesday, a Catholic Mass was scheduled right after dinner for anyone who wanted to go. Amid all the turmoil and indoctrination in those three days, Mass felt so familiar. I was immediately grounded. I was too dazed to pray for help, but help was there, nonetheless. I never felt ill again for the duration of Swab Summer.

We ran, did calisthenics, double-timed, braced up against the wall, marched, and tried hard to memorize all kinds of useful and useless information. We swam, ran the obstacle course, learned how to properly tuck in a shirt, polish shoes, and change from one uniform to another faster than we ever thought possible. And they tried to teach us how to sail in small sailboats. They were only slightly successful in my case.

For the first time, we were each held rigidly responsible for everything we did, how we looked, and what we had to memorize. We were also somehow responsible for each other as well and shared the repercussions of a classmate making a mistake or not knowing something. I did more pushups than I ever thought possible, and as a group, we tried hard to get to eighty-two pushups while our cadre from the Class of 1980 stopped at eighty pushups. In the process, we became a tightly knit group: the Class of 1982.

Swab Summer was hard, harder than anything I had done before. It didn't help that my roommate, a kid from the Florida panhandle, dropped and broke a full bottle of Musk cologne in our room shortly before I got to the room. He had cleaned up most of it, but the room smelled of Musk the whole summer. That smell still nauseates me to this day.

Near the end of the summer program, we were flown to Seattle and sailed the Coast Guard Cutter EAGLE from there to San Francisco. Onboard, we had no clue what we were doing, but the officers, enlisted crew, and 2nd class cadets (cadets getting ready to start their junior year and also serving as our training cadre) trained us on the fly and kept us from serious injury. We learned a lot of basic seamanship and how to sail a 295-foot, three-masted barque, a classic type of old sailing ship with square sails on the forward two masts and fore-and-aft sails on the aft mast. This Coast Guard training ship was a throwback to days of old, but it taught us lessons about the sea, the weather, and ourselves that were as valid as ever. We pulled on more ropes with more odd names that did strange things. Sometimes, we were just taking up the slack in lines, but other times, we were pulling hard. And we stood watches: helmsman, lookout, ready boat crew, and engine room watch.

We had training lectures and day work, all mixed in with unannounced "sail stations" where we tacked to change course with commands cascading down from the bridge to each mast in turn. The berthing areas were tight with beds stacked three-high and limited locker space for clothing and gear. The food was mediocre at best. I thought I had gotten a small steak one day for lunch, but it turned out to be a piece of liver. I only made that mistake once. And the smell from the scullery, where the dirty dishes and pots were washed, was horrible; hot steam mixed with food scraps. And we discovered that some of our classmates got seasick. I may not have felt well on occasion, but I proudly never yacked onboard EAGLE, and certainly not in the berthing area or on the mess deck or into the wind. Yep, we learned a lot about basic seamanship.

Back at the Academy, after our training cruise, there were informal donut socials after Mass every Sunday outside next to the chapel. It was supposed to be a bit relaxed. They had us remove our shoulder boards indicating our rank (or lack of rank for us new 4th Class cadets), but we all knew who were "swabs" and who were training cadre. The donuts were much appreciated, and the Catholic Chaplain, Father Keefe, was great.

With September came the start of classes and the start of football practice. I didn't even come close to making the varsity squad, but football practice brought a sense of normalcy to the still-new routine. The academic year was challenging, mostly because of the number of classes on top of all of the other 4th class cadet restrictions and responsibilities. Our days started early and often with marching practice. Classes each morning were followed by a uniform inspection just before lunch, then classes in the afternoon until 3:30ish. Participation in intercollegiate or intramural athletics was mandatory for two of the three sports seasons, and practices or intramural games were always late afternoons. Another uniform inspection preceded the evening meal, which was followed by study hour until 10:00 pm. Then we had a fifteen-minute "break," which for 4th class cadets involved being "braced up," i.e., standing rigidly at attention with your chin tucked in against the wall in the corridor and being lectured or yelled at about everything that had been done wrong and what we needed to do better. 10:15 was taps and lights out, although you could stay up studying as late as you wanted. On top of that, we had to always know the menus for the next three meals, plus the upcoming opponents of each Coast Guard Academy sports team, plus a whole lot of Coast Guard indoctrination material such as the chain of command from the cadet level all the way to the Commandant of the Coast Guard.

We were rarely told how to get a job done and were expected to figure it out. This was referred to as using our "trained initiative." Whether you learned to shine shoes and polish your brass belt buckle by learning from a classmate or by trial and error, you figured it out. The only appropriate response to a question about why your shoes weren't shiny enough, or your tie was loose, or your room was not spotlessly clean and meticulously organized or "squared away" in Coast Guard language was, "No excuse, sir." Most of us got by on six hours of sleep, and we all stood accountable for ourselves and our mutual performance in every respect. It was exhausting and exhilarating at the same time.

Weekends offered only some respite. Saturday mornings involved room inspections and inspections of the common areas. We worked fast and hard to make everything just right, then stood in front of our bunks while one of the company officers or 1st class cadets did the inspection. Sometimes, it was very thorough; other times, it was a cursory inspection. You never knew. Once, the 1st class cadet spent five full minutes inspecting my roommate's half of our room and ten seconds on my half of the room. He suspected my roommate of wrongdoing of some sort and was determined to find something so he could punish him. He tore his side of the tiny room apart but couldn't find anything. My roommate and I just stood there and patiently waited for him to finish. That 1st class cadet never did find anything.

The Corps of Cadets was organized as a regiment with four battalions of two companies in each battalion. Each company had cadets from the freshman class through the senior class. I was in Delta

Company. Periodically, my classmates in Delta Company would conduct a "skit night" during the 10:00-10:15 pm "break" at the company level that involved crude, homemade comedy skits. Our target audience was the group of 1st class cadets from the Class of 1979. They were the last all-male class at the Academy since the 2nd class cadets from the Class of 1980 included the first women at the Academy.

Cadets from the Class of 1980 were our cadre and were responsible for training us idiot 4th class cadets. Skit nights were a chance to poke fun at anyone and everyone. The cruder the skit, the more the 1st class cadets seemed to like it. I had neither the creativity nor the time to devote to crafting ideas and scripts, but I had several classmates who excelled in that area, and I could always be counted on to participate.

One night, a few of the skits were particularly raunchy and seriously poked fun at several of the female 2nd class cadets and their classmate boyfriends. They got pissed and made all of us 4th class cadets in the company report to one of their rooms. We were all braced up while they yelled about how much trouble we were in. They wanted to know who had authored the most egregious of the skits. None of us admitted to it. The truth was that I had no idea who wrote it. I had been studying and was only brought in right before the skit to learn my part. The angry 2nd class cadets went around and asked each of us if we had written it. We all answered negatively, which sent them into a frenzy. Lying was an honor violation and would be grounds for dismissal from the Academy.

Things were getting serious when there was a knock on the door. It was one of the 1st class cadets informing all of us 4th class cadets that the company commander (a 1st class cadet) wanted to see us. As it turned out, my classmate who wrote the offensive skit was in his room, the very last room at the end of a long hallway. He never got the word that he had to report around to the angry 2nd class cadets with the rest of us. He was sitting in his room studying while we were getting yelled at. He would have gladly joined us had he known about it and would have admitted authorship, but he wasn't there, and the few who knew were not going to rat out a classmate.

When the 2nd class cadets were done with us, we dutifully trooped down the hallway to the company commander's room and piled in. It was a "room stuff," made all the more claustrophobic by the two thick cigars the two 1st class cadets had been smoking and the cloud of thick smoke that filled the upper half of the room. The company commander started by saying, "First off, it was a fine skit night." After that, we knew we would be fine. None of us ended up with demerits, but the angry 2nd class cadets all got demerits for "creating a disturbance after taps." Life isn't fair sometimes.

At the end of the spring semester and before graduation events took place, there was a one-week scheduled lull in the activities. We were to be shipped out right after the graduation on two, five-week training cruises, one on the EAGLE again and one on an operational Coast Guard Cutter. Before that lull, the Academy asked for volunteers to take the EAGLE out on a shakedown cruise to prepare it for the longer summer cruises. I was thinking about volunteering when a close friend, Bruce Gaudette, informed me that he had already volunteered the both of us. I could have backed out, but decided to go for it.

I took the Coast Guard option again, and what an experience. Instead of a hundred and fifty cadets onboard for the long cruises, there were only about forty of us. We worked hard all day long and stood watches as well. We did all kinds of maintenance, from greasing pulleys at the ends of the yardarms to replacing old and worn ratlines used to climb aloft. The good part was that we were not subjected to boring training lectures or other such nonsense. We just worked. We tested every line, every ratline, every sail. We launched the small boats, a very manual effort on the EAGLE involving pulleys and ropes and a lot of coordination. And we pulled hard on so many lines that, for most of us, our fingerprints disappeared because of working with the wet, manila lines.

And the weather sucked. It was early May off the New England coast, and cold, rainy, and foggy. One night, I was standing watch as the stern lookout somewhere between midnight and 4:00 am. I was wearing every layer I had brought with me underneath my yellow foul weather gear along with binoculars and a sound-powered phone so I could communicate with the bridge in the event I saw anything. But I couldn't see anything beyond about twenty yards due to the fog. At some point, one of the officers onboard, a Lieutenant who had been my coastal navigation instructor the previous year, made a round on deck and saw me standing there. He started laughing because he saw the futility of having a stern lookout when visibility was so limited. "Mr. Close," he said, laughing, "are you learning anything?" "Yes, sir," I replied, "I'm learning not to volunteer." He laughed even harder, then left me to stand my watch with just the fog as company, the frequent blare of the foghorn, and an occasional porpoise playing in the ship's wake that caught just enough of the stern light to be visible. It was funny, but not a lesson that I truly learned, as I would continue to volunteer in the future. There were too many things to see and experience to just sit back and not participate.

With the minimal crew onboard, it was as close to sailing in the old days as we would ever get. Almost every minute of that shakedown cruise sucked, but I never regretted it.

After two weeks back at the Academy for the graduation events and parades, the formal summer training started. I was back aboard the EAGLE for the first five weeks. We sailed to Halifax, Nova Scotia, and then to Alexandria, Virginia. It was very different from the shakedown cruise with so many cadets onboard, but at least we knew what all of the numerous ropes did, when to pull hard, and when to let out slack in the lines. The port visits were nice, but the weather in between was wet and dreary.

Later that summer, I was onboard the Coast Guard Cutter Ingham, an old 327-foot Medium Endurance Cutter that had been built in 1936 and had guarded North Atlantic convoys during World War II. She was an old steam-driven ship that rode well and had good bones, albeit tired bones. Some of the regular crew had been temporarily transferred off the ship to make room for us cadets, a mix of upcoming 1st class cadets (seniors) and us upcoming 3rd class cadets (sophomores). We reported onboard and stowed our gear as the cutter headed south for the Caribbean and better weather. I recall that INGHAM mostly operated independently from other Coast Guard cutters, and we conducted training while on a law enforcement patrol looking for drug smugglers. It was an introduction to how big the ocean was.

Once again, we stood watches, suffered through dry training lectures, wolfed down insufficient quantities of mediocre food, and tried to find ways to catch up on sleep. As 3rd class cadets, we stood watches that were normally stood by junior enlisted persons: helmsman, lookout, messenger, and engine room watches.

Eventually, INGHAM stumbled across a suspicious-looking shrimp boat way south in the Caribbean and heading north. When he saw us, he immediately turned back to the south. Since the vessel was flying the flag of one of the Central American countries, we had to get their permission to board the vessel. It took a couple of days for the flag state to declare they had no such vessel registered. That pronouncement made the vessel "stateless" and gave us all the authority needed to board it. We temporarily lost the vessel in a thunderstorm, and when we found it again, it was still heading south and only a few miles from Venezuelan waters. We went to "general quarters," and the active-duty crew (no cadets) geared up and prepared to launch the INGHAM's port side small boat to conduct a law enforcement boarding.

While that was happening, the INGHAM approached the shrimper and, over both the radio and loud hailer, ordered it to stop in English and Spanish. The shrimper responded by turning into the INGHAM and ramming us. She t-boned us about midship, on the starboard (or right) side. Ironically, it was the age of the INGHAM that prevented any real damage. She was built at a time when the hull plating and side shell were thicker than more modern ships, which derived their hull strength from the framing and not from the hull thickness. The shrimper only scraped some paint off the side of the INGHAM's hull, but as the shrimper turned, one of her outriggers damaged railings on the main deck and the deck above, exactly opposite where the crew was getting ready to launch the small boat on the port (or left) side.

I was already at my assigned general quarters station in the Combat Information Center located in a space behind the bridge. Shortly after getting rammed, I heard and felt the unmistakable sound of a .50 caliber machine gun firing a few three-shot bursts. The shrimper was circling and lining up for another ramming attempt until a few shots were fired across their bow. That stopped them immediately.

The boarding went smoothly from that point, and the shrimper was quickly found to be full of what turned out to be seventeen tons of marijuana in bales. The drug runners were cuffed and transferred to the INGHAM without incident, but with a lot of dirty looks from INGHAM crewmembers who did not at all appreciate being rammed. The drug runners had disabled the engine on the shrimper, so we took the vessel in tow and placed a few people onboard as "prize crew" to keep her floating.

Early the next day, at a training lecture, one of the Academy officers onboard INGHAM asked for volunteers to replace the prize crew. He said only people who don't get seasick should volunteer. I raised my hand anyway. It just sounded like too much of an adventure to pass up. I was all in and took the Coast Guard option again.

As luck would have it, the Academy officer had been my calculus instructor and knew me, so I got picked. Boarding the shrimper while being towed was a bit tricky. The seas were choppy, and the INGHAM's small boat never was in sync with the rising and falling of the shrimper. It had to be timed just right, but I made it onboard smoothly. One of the enlisted guys was next in line to come aboard. He was a good guy, but thick around the middle and not very athletic. He stood up on the small boat's gunwale and grabbed the railing on the shrimper just as the small boat started to drop between waves. I was leaning over the rail watching, and when I saw the small boat start to drop, I grabbed the guy and pulled hard to prevent him from hanging there and maybe falling between the boats. I was successful, but as soon as I pulled hard, the shrimper dropped between waves, and the guy flew onto the shrimper and landed right on top of me. He pancaked me pretty good, but no one was hurt.

Duty as prize crew turned out to be pretty boring, and we only had to contend with the sickly sweet smell of marijuana that was fermenting in salt water. Even though the shrimper only scraped paint off the INGHAM, the bow of the shrimper was crumpled and split open enough to allow some saltwater spray to enter as it was being towed. It was tolerable, but mostly, we stayed in the fresh air. Then, in the middle of the night, the weather picked up, and the shrimper started crashing down into the waves. A wave blew out one of the forward windows, sending a mass of green water into the pilot house and soaking us wide awake. While one of the enlisted crew got on the radio to tell the INGHAM to slow down, the rest of us got busy shoring up the open window to keep more water from coming in. Even with limited tools, we managed to break up a bench seat and secure some boards over the window. It wasn't pretty, but it worked well enough. The next morning, we were swapped out for another prize crew. I was exhausted but glad I had volunteered.

The rest of the cruise was uneventful, and we all were looking forward to three weeks off before reporting back to the Academy. It was an odd visit home. The rest and good food were most welcome, but I was different in ways that none of my old friends from high school could relate to. Their college stories were about parties and drinking. They just would not be able to understand the rigors and variety of my experience, nor the bonds with my classmates formed by shared difficulties and challenges. I was correct in believing that my choice of college would impact the rest of my life; I just didn't know in what ways at the time. But I was starting to understand.

Back at the Academy for my 3rd class cadet (sophomore) year, things were definitely better. We still had to meet all the same standards, but were not nearly under the scrutiny that we were as 4th class cadets. There was a whole new group of 4ᵗʰ class cadets, and we were seasoned 3ʳᵈ class cadets by then. Academics were finally starting to relate more to my major. I was still listed as a Marine Engineering major because I had no knowledge of the field yet, and thus had no reason to change majors. I decided to wait and see once the courses got more major-specific in my junior year before making any changes.

The following summer, as an upcoming 2ⁿᵈ class cadet, was very different. We were sent all over the country for one to two week "training" opportunities. A good friend, Mark Butt from Iowa, and I spent two weeks on an 82' Coast Guard patrol boat. I spent a week in Orono, Maine, as a counselor at the Maine Boys State program. A group of us also spent a week in Philadelphia at a Navy training facility learning about ship damage control. We also got cycled through a two-week program in Mobile, Alabama, at the Coast Guard Aviation Training Center. We suffered through numerous lectures before they took us up for rides in Coast Guard helicopters and C-130s. We also got to experience dunker training, simulating a helicopter crash that drops into the water and rolls over. The training was to try to get out alive. That accompanied some swimming training in a pool in a flight suit and boots, and survival training in Mobile Bay with large, inflated life rafts and helicopter hoist lifts. The most educational few weeks that summer turned out to be back at the Academy as cadre for the incoming cadets. Leadership is best learned by doing, and for many of us, it was the first time we actually felt in charge of a group of people. It was different than on the football field, different from being an officer in a high school club, and different from informal leadership with friends. I welcomed the opportunity and felt surprisingly comfortable in a leadership position. It felt good.

That was also an opportunity to learn that training new cadets and holding them accountable for everything they did and did not do, did not mean you had to be a prick about it. You weren't looking for friends, but you could be fair and professional and achieve results.

At the end of those two weeks, it was traditional for the new 4th class cadets to attempt to toss the 2nd class cadet cadre into the showers or some similar prank kind of thing. Of course, it was the job of the 2nd class cadet cadre to do everything they could to thwart those attempts. Since I had duty that night and had to sleep in the duty room near the front door of the barracks, I wandered up late in the evening to get my overnight things. I heard some commotion from down the hall where Mark Butt was cadre for a platoon of swabs. One quick look and I knew what was happening: they had grabbed their three cadre, including Mark. This could be fun, I thought. Since I was still in full uniform, I exercised my Coast Guard option, brazenly marched down the hall, and yelled for all the swabs to brace up against the wall. They outnumbered me thirty to one, but I projected authority, and they didn't know who I was. They hesitated for a second, then jumped to attention with their backs against the walls, and I owned them from that point.

Mark and one of the other cadre (also my classmate) had been hog-tied and doused with shaving cream. The hallway was a mess. With the most serious face and voice I could muster, I yelled at the swabs for their behavior and told them they had three minutes to get everything cleaned up and get in their rooms. With that, I stormed away for a few minutes and hoped they did not see the smirk on my face as I tried hard to keep from laughing. Sure enough, when I returned, the place was immaculate. The floors and walls had been wet-mopped, and there was no sign of the commotion from earlier. I loudly marched down the hall, and you couldn't hear a sound. Until Mark came out and quietly asked me if I was really mad. I almost started laughing but stayed in character just long enough to clear the area. Out of the corner of my mouth, I whispered to Mark that he sure did look pretty funny, all tied up. Somewhere in there was a lesson about faking it until you make it, or maybe just acting like you are in charge is enough to make others believe you are in charge.

Classes in my junior year finally started to focus on marine engineering topics. The more I learned, the more interesting it became, especially ship stability and ship propulsion topics. Football got better, too, as I was playing consistently and having fun. It wasn't a good team, but it was a good group of guys. I found academics to be easier during football season for some reason, maybe because there was less time to fool around and fewer distractions. I also found that I had a steady stream of classmates who came to me for help with homework or understanding academic concepts. I always helped, but was sometimes frustrated that it ate into what little free time I had.

Tragedy struck in the spring when a classmate, Julie Babineau, was killed in a freak accident during sailing practice on one of the Academy's ocean racing yachts. There was some miscommunication as the sailboat turned, and the boom swung quickly across the boat and struck her hard. Julie was a good sailor, a good person, and a beautiful girl who would have made a fine Coast Guard Officer. Her death hit many of us hard. It was a reminder of the fragility of life and that the risks were real. There was no grief counseling or anything remotely similar at that time. We leaned on each other. Life would go on.

Our final summer as cadets was again divided into two, five-week programs. Mark Butt and I spent the first five weeks on the Coast Guard Cutter CONIFER, a 180' buoy tender out of Atlantic Beach, North Carolina. We were the only cadets onboard, and training consisted of getting things signed off in a "Professional Qualifications Standard" or PQS book. Mostly, we stood navigation watches underway, and officer of the deck watches in port while helping the ship get caught up on routine stuff no one else had time for.

During the second half of the summer, Mark and I were back on the CGC INGHAM. We met the ship in Rota, Spain, and made stops in Amsterdam and Kristiansand, Norway, before a refueling stop in Newcastle, England. Then it was back across the Atlantic at a slow bell so we could suffer through more training lectures and stand a lot more watches. By that point, most of us knew whether we were planning on being engineering officers or deck officers when we graduated, and the ship was smart enough to let us focus on one or the other while onboard.

That old steam plant was a marvel; minimal automation required watchstanders to stay alert and focus on dozens of gauges and water-level sight glasses. It was an old school, no air-conditioned control room. Watches were stood in the heat, humidity, and noise of the engine room. With training over, we had three weeks off before reporting back to CGA. By this time, I wasn't even looking up friends from high school. We had nothing in common any longer. I spent some time with my grandparents in Pittsburgh and at home with my folks, and tried to get in shape for football after a summer with limited opportunity for exercise (or sleep) or good food.

It was much better being a 1st Class cadet (senior). We had cadet positions within the platoon, company, and regimental structure, and responsibilities that went along with them. The positions rotated three times during the academic year to give more of us experience in the different positions. Most visibly, your position was obvious during Friday afternoon parades in the fall on the Academy's parade ground. They were admittedly impressive, with a thousand cadets in dress uniform with rifles for the underclasses and swords for the 1st class cadets. The band would play, flags would fly, and Academy officials or the occasional dignitary would troop the line. Bayonets would flash as they were fixed on rifles as part of the parade, and cadets would march off smartly. Cadet companies would be graded by officers, and the best company would be awarded points.

I was privileged to be selected as the Regimental Commander for the second rotation. It was an awesome thing to march out onto the field, have the entire corps of cadets parade past, then lead them off the field. Beyond that, the position mostly involved daily meetings with the Assistant Commandant of Cadets and two of the commissioned officers who oversaw the companies. They would go over upcoming scheduled events and list things that cadets had done wrong and which cadets received demerits, while I jotted down a few notes.

My academic classes were concentrated on my major, but we still had other required classes to make us "well-rounded" and prepare us to be Coast Guard officers. My classmates and I in the engineering majors moaned about being forced to take a management class, a psychology class, a history class, and navigation classes, just to name a few. We could not get enough engineering classes to satisfy our desires, but our expectations were probably skewed a bit. Similarly, the Government majors moaned about being forced to take a class titled, "Basic Naval Architecture."

The faculty was a combination of Coast Guard officers who were assigned temporarily or permanently to the Academy to teach, plus some civilian professors. At the suggestion of one of the civilian faculty members, I applied for a Hertz Foundation Fellowship at the end of the fall semester. I had never heard of the Fellowship or the Hertz Foundation, but the Fellowship was arguably akin to a Fulbright or Rhodes Scholarship, only for engineering disciplines. I don't remember what I wrote, but it was sufficient to move me to the next round, a technical interview in New York City with someone from the Foundation.

I was nervous about the interview and had absolutely no idea what to expect. I took the train to NYC, looking sharp but very much out of place in my cadet uniform. I grabbed lunch and still had some time to kill before my scheduled interview. I wisely spent part of it in Saint Patrick's Cathedral, which was impressive and quite beautiful. A few prayers didn't hurt either. The interview consisted of a series of generic science questions, and I felt like I bombed. The only saving grace was that the interviewer's father had, at one time, been in the Coast Guard, and the interviewer recognized that he knew nothing about marine engineering or naval architecture. He determined that I should get a do-over with someone more knowledgeable in my field. So, a few weeks later, I repeated the process.

I was to meet the new interviewer in his hotel room in New York, which sounds a little creepy today, but less so at the time. He had flown in on the redeye from California, where the Foundation was headquartered and would be flying back right after the interview. I called him from the lobby when I arrived at the hotel to let him know I was there and confirm the interview time and his room number. At the appointed time, I knocked on his door and heard nothing. I tried knocking louder and then heard snoring. It seemed he was taking a nap after the redeye flight. Wondering just a little whether this was part of the interview, I wandered back to the elevator lobby and used the house phone to call his room. After many rings, he answered. I lied and told him I was in the hotel lobby so he wouldn't know I was basically standing twenty feet from his door. I sat in a chair in the elevator lobby for the longest five minutes of my life, then walked the short distance down the hall and knocked again. This time he answered, albeit looking a bit disheveled from his nap.

I don't recall what his field was, but it wasn't marine engineering or naval architecture. However, he had apparently studied up on the flight out, an impressive thing and an indication of the kind of smart people associated with the Foundation. We had a good discussion about ships and ship design, and I shared some information from a professional society lecture I had heard a few months previously about hull vibration caused by the ship's propulsion system and design approaches to minimize vibration issues. He was suitably impressed, and later that school year, I got a phone call congratulating me on being awarded a Fellowship, the first ever for anyone from the Coast Guard Academy. I was awestruck and had no idea how that would work with the Coast Guard, but was confident we'd figure it out.

At some point early in the spring semester, one of my football teammates found himself in trouble and at risk of getting thrown out of the Academy. More significant than being a teammate, Bob was a classmate, and the Academy had spent years drilling into us that it was all about your class and your classmates. You had to be a good classmate. You helped your classmates. You never screwed a classmate. From the very first day, if a classmate got in trouble, you rushed to join him or her and share the punishment. It was what classmates did. And we are all still tight decades later, as is each class that came before us and each class that came after us.

At the end of each semester, cadets were evaluated by their peers and their company officers (lieutenants or lieutenant commanders). As 4th class cadets, the evaluations were mostly done by other cadets who would rank order all other cadets within their class. As 1st class cadets, the evaluations were heavily based on the company officer's evaluation. Cadets who were evaluated near the bottom were brought before two review boards that would decide whether they still had potential or should be tossed. Bob was never a very squared-away cadet, and his academic performance started off pretty rocky. He had been in front of these boards once before as a 4th class cadet and had pretty easily convinced them that he was worth keeping. This time, he took it a bit too lightly and had not thought through what he would say. They recommended that he be dismissed from the Academy, but allowed him to appeal the decision. Having learned of this, the head football coach met with me and another 1st class cadet on the football team, my friend Mark Butt. Coach had been talking with the Athletic Director, Otto Graham, who was famous for his professional football career and a bit of a legend around the Coast Guard Academy. He also held the honorary rank of Captain. Our football coach asked us to meet with Otto Graham and discuss if there was anything we could do to keep Bob from getting the boot. Otto Graham said that if we tried to do what we could to help Bob from our end, he would work the senior officers at the Academy from his end. It sounded like a plan, although a vague one.

Mark and I put our heads together and dug through the Cadet Regulations to see if there was any loophole or approach that hadn't yet been explored. In the process, we reviewed Bob's record with his permission. While Bob had been on academic probation as a 4th class cadet, his grades had improved each year to the point that he had just made the Dean's List for the fall semester. His evaluation scores from our classmates for the fall semester had improved to being solidly in the middle. That was nothing to write home about, but didn't seem to merit going before the review board.

Impressively, the evaluation he had received from the Commanding Officer of a Coast Guard patrol boat he trained on the previous summer was outstanding. He noted that Bob had quickly qualified to stand watches underway and that Bob was the only person onboard that he trusted to stand watch at sea during the time period that his Executive Petty Officer was off the boat due to a family medical emergency. He indicated the boat would not have been able to get underway to respond to a case had Bob not been onboard. It seemed to us that Bob just needed to get away from the artificiality of the Academy and get busy doing real Coast Guard work. He ended up in front of the review board mostly because his company officer rated him poorly, and that counted for about 85% of his evaluation. It just seemed wrong that one lieutenant's evaluation of Bob had so much influence.

We came up with a scheme and gave it a shot. We were exercising the Coast Guard option in the form of some trained initiative. We were going to seriously buck the system, but more so, it was the right thing to do. It was fighting the good fight. While petitions were specifically prohibited for cadets, we asked each of our classmates to sign an individual letter to the Superintendent, the two-star admiral in charge of the entire Academy, asking that Bob's appeal be favorably considered. Of our one hundred and fifty-four classmates, we collected and submitted roughly one hundred and forty signed letters. Then the shit storm arrived.

Mark and I felt bulletproof. I had been the Regimental Commander, and he had held a similarly high position. We had both done very well academically, were ranked near the top of the class, and had pretty stellar records. Nonetheless, my company officer was pissed. Another company officer who had been an assistant football coach of mine previously was furious with me. He cornered me and screamed until he was red in the face. It still seemed that Bob was getting screwed, so, we stood by our classmate. I took the Coast Guard option and held my ground. It still seemed like the right thing to do, and it seemed like it was in the Coast Guard's best interest.

Eventually, the shitstorm subsided a bit. At my request, I got an audience with the admiral, which was extremely uncommon. He allowed me to say my piece, and I was convinced that he at least considered the argument. In the end, Bob was allowed to graduate with his degree but not with a commission as an ensign. He was shipped off as an enlisted man, a Quartermaster First Class, to the Coast Guard Group in San Juan, Puerto Rico. His qualification on the patrol boat from the previous summer made him a sought-after commodity with the patrol boats based in San Juan, and he excelled doing real Coast Guard work. Bob was eventually commissioned and served honorably for many years. We had the last laugh, and the Coast Guard benefited. Classmates are for life.

In those days, cadets picked their first assignments. Headquarters provided a list of all the billets coming available on various Coast Guard cutters, and we picked our billets in order of class rank. The Hertz Foundation had agreed to defer the start of my Fellowship until my two-year assignment on my first cutter was complete.

"Billet Night" was a grand night that we had anticipated for years, and there was much celebrating. Beer was consumed by the class-mug full. My good friend, Bruce Gaudette, and I selected the US Coast Guard Cutter STEADFAST out of St. Petersburg, Florida, and were joined by another classmate, Dave Sella. I was to be an assistant engineering officer, while Bruce and Dave would be deck officers. It all started to feel especially real at that point.

Graduation week was a whirlwind of activities. My family was there on a beautiful day where degrees were conferred and ensign commissions presented. With a big yell, we tossed our cadet hats into the air and donned our new hats as commissioned officers.

We had a few weeks off during which time I broke off an engagement to a great girl that I hardly knew, who got cold feet. We had no business getting engaged, but I had gotten swept up in the romance of "graduating from college and getting engaged" as several of my classmates did. I was okay with ending it. She had been a middle school sweetheart from Waynesburg whom I had corresponded with when we first moved to eastern Pennsylvania, then lost contact with, only to rekindle the letter writing while I was at the Academy. It was a romantic notion, but I didn't really know her well at that point. I walked away silently, wishing her well.

Then it was off to St. Petersburg and Coast Guard Cutter STEADFAST (WMEC-623).

THREE

The Start of Active Duty (1982-1984)

S TEADFAST was a 210-foot Medium Endurance Cutter built in the late 1960s in Lorrain, Ohio. 210s were unique in a few ways. To accommodate a flight deck for helicopters and to make the Cutters harder to see from a distance, the exhaust stacks ran horizontally from the engine room aft and out the stern. The theory was that the exhaust from the 210's engines would be less visible than on most ships that sent a plume of exhaust skyward. With 210s, the exhaust would dissipate close to the water.

210s were also known for how they rode in rough seas: not well. They were a bit top heavy and had a sloping bottom at the stern so that in a following sea, they would roll quite a bit. At least being located in Saint Petersburg and normally patrolling along the drug smuggling routes in the Caribbean and the Gulf of Mexico meant better weather than the 210s home based up north. To help the ship ride better, the fuel storage tanks could be filled with ballast water when necessary, but that came with its own problems. When discharged, the ballast water could be contaminated with some residual fuel. But in 1982, that was not seen as the pollution problem it poses today. Conversely, when more fuel was added, the residual water would have to be filtered out and sometimes the layer between the fuel and the water would allow algae to grow that could clog up our fuel filters. I learned quickly that on STEADFAST, we only ballasted when absolutely necessary and lived with the rougher ride.

Bruce Gaudette, Dave Sella, and I met STEADFAST while the Cutter was on a mid-patrol break in Key West, and we quickly jumped into our duties. My job was to learn the engineering plant, including tracing out piping systems and learning which valves did what. I needed to know that to qualify as a watchstander, beginning with the most junior watches in the engine room, usually assigned to junior enlisted crewmembers. Once qualified at that level, I was allowed to begin learning the watch position usually assigned to more senior enlisted crewmembers. Only when I was qualified and experienced in that position could I break in and qualify as the Engineer of the Watch (EOW). The EOW was responsible for the machinery and systems, as well as the two more junior watch standers on watch.

I rolled up my sleeves and worked hard. The crew was a mixed group of talent, and the plant was not in terrific condition, but I worked well with all of them and developed a respect for the enlisted men who usually had longer tours afloat and did all of the grunt work.

Most of the STEADFAST's patrols were counter-drug patrols in the Caribbean, or near the Bahamas, or between Cuba and Mexico. On occasion, we found ourselves off the coast of Haiti picking up Haitians who were in small, unseaworthy boats trying to sail to the U.S. It was sad duty bringing them back to Haiti even though we knew we were probably saving their lives since they had very limited food and water and little or no experience navigating. Most had pooled all of their money for the boat that came with navigation directions as specific as "follow the sun." While we were saving them, we knew we were returning them to Port-au-Prince, Haiti, and the same hopeless squalor they came from.

As a Medium Endurance Cutter, our food and fuel capacity was limited, so every ten to twelve days we had to pull in for a patrol break, usually in Key West or Guantanamo Bay (GITMO), both of which got old quickly.

My duties were well-defined except in one area. When I reported onboard, I was appointed the ship's morale officer, complete with a small budget, big expectations, and the keys to the unit's soda machine. I held that collateral duty for the entire time I was onboard. Besides keeping the soda machine stocked, I was responsible for some underway morale events and some sort of party during our mid-patrol breaks in GITMO. I also managed some sports equipment and some fishing gear for when we had the occasional "fish call" while underway.

The mid-patrol breaks at GITMO usually involved coordination with the cooks to grill burgers and dogs at a park on base, while the morale fund provided the beer. The crew became convinced that the beer we purchased at GITMO contained a special preservative to enable it to handle the long, hot trip by barge from the U.S. mainland to GITMO on the southern coast of Cuba. They maintained that the preservative gave them headaches. I always suspected that their headaches were more of a hangover from the free beer they consumed. Nonetheless, the Commanding Officer agreed that we could bring our own beer for one trip, provided I had a locked space to store it and that he and I were the only ones with keys. Thus, the night before we got underway, I was at the local beer store buying numerous cases of beer. The clerk asked if we were having a big party, and I figured I'd mess with him and just said, "Nope, we're going to Cuba." If memory serves, the mid-patrol break was a success, and no one admitted to headaches, just hangovers!

Overall, we sailed a lot of miles on STEADFAST and got to see a few interesting places, such as the events associated with Saint Kitts and Nevis being granted their independence from the United Kingdom. Still, we were not very successful in terms of drug busts. One thing we were good at was finding poor Haitians trying to sail to the U.S. on some simple vessels that lacked basic navigation equipment. Many of the Haitians were unhappy with us for taking them back, but begrudgingly grateful we had found them. After bringing them onboard, we destroyed the old, wooden, leaky sailboats.

We made several trips into Port au Prince, but it was not much of a liberty port. Inside the harbor, young men and boys would swim or paddle to the STEADFAST, trying to sell or barter wood carvings. That was dangerous for them since we were still underway at the time, but their efforts reflected their level of desperation. Haiti was a third-world country led by a dictator at that time. Everything was rundown and in a state of disrepair, with one exception. The Presidential Palace was beautiful and well-maintained, but we were warned not to walk near the fence that surrounded it, as the guards would shoot. Lovely. A few of us walked around a bit during daylight and bought a few small things, but only once.

The STEADFAST was a valuable experience for me as an engineer because I got to see a lot of repairs. I was generally unimpressed with the Coast Guard's support for the cutter's engineering systems and engineers. For example, we always felt unfairly judged as to why our engine room didn't shine to the same extent as the engine room on sister cutters based out of Miami. But the cutters homeported in Miami had the advantage of a large shoreside maintenance detachment that was ready, willing, and able to assist with maintenance and repairs and special projects during their in-port periods. We had no such luxury. Plus, the previous Commanding Officer was the type that knew only one speed, so both main engines always ran at nearly full power, which made the engine room dangerously hot in the Caribbean and caused all maintenance to be deferred to the in-port periods that were already busy enough with repairs. Despite being able to keep everything running, we had little time or budget to clean years of soot leaks and apply fresh coats of paint.

And there seemed to be no simple solution. Even the arrival of a new Engineering Officer (my boss), a prior-enlisted lieutenant with years of Coast Guard engineering experience, saw no major improvement as we were caught in the crisis management loop of working our butts off, barely keeping everything repaired, and keeping the cutter mission-capable without the chance to fix things properly so that they didn't need crisis repairs. It was a most educational experience, if somewhat disheartening.

The real highlight of the tour was that I got married in March of 1984. Peggy O'Keefe lived in Allentown, Pennsylvania, and I met her for the first time when I was a 3rd class cadet when she was thinking about attending the Coast Guard Academy. Her elderly neighbor's grandson knew my girlfriend at the time, and we made the connection and spent an hour on the phone talking about the Academy. We met a few months later at the Academy when she came up for a visit. I recognized her at lunch because of the bright Allen High School jacket she was wearing (a rival high school of mine).

She entered the Academy as a 4th class cadet in the summer of 1980, and my classmates and I were the training cadre for her class. Consistent with Academy rules, fraternization with 4th class cadets was prohibited, and I respected that. But we did share a few train rides back to Pennsylvania. In the fall of her 3rd class year, she decided that the Academy and the Coast Guard weren't for her. I respected that decision and agreed that there were few good female role models at the time. Plus, she had recognized that her real calling was physical therapy, a far cry from what the Coast Guard had to offer. We reconnected right after my graduation and stayed in touch. She visited me in St. Petersburg a few times, we fell in love, and I proposed.

We were married in Allentown at the Cathedral of Saint Catharine of Siena. I already knew that I would be getting orders to start graduate school in Boston at M.I.T. in a few months when I completed my tour on STEADFAST, but Peggy's physical therapy graduate school plans were still uncertain as the wedding day approached. As Peggy's father walked her down the aisle and to the altar, she asked the priest if she could tell me something first. The poor priest had a somewhat shocked look on his face, but said yes, of course. She told me that she had just received her acceptance letter for the two-year

Physical Therapy program at Boston University starting that summer. We would both be in graduate school at the same time and in the same city. It was meant to be, and it was a great wedding!

Our honeymoon in early March was in Boston, where winter was still in full force. With us both locked into a move to Boston, we used our honeymoon to find a place to move into a few months later. Right after our honeymoon, Peggy went back to her folks' house in Allentown and finished her last semester at Muhlenberg College while I went back to St. Petersburg and got underway on another patrol on STEADFAST. Welcome to the Coast Guard, Mrs. Close.

FOUR

Graduate School (1984-1986)

After two years on STEADFAST, I was officially transferred and started graduate school at M.I.T. with the deferred Hertz Foundation Fellowship. The Coast Guard regularly sends junior officers to graduate school for a master's degree, so it was no surprise that I wasn't the only Coast Guard officer there. The great surprise was that another classmate of mine, John Kaplan, was also reporting in. We were both sponsored by the Coast Guard's Naval Engineering program that oversaw engineering on Coast Guard cutters and boats, and we both enrolled in the Naval Architecture program at M.I.T.

A few days before classes began, we were invited to attend a meeting with the U.S. Navy officers in the same program. Not only did the Navy send several officers each year to M.I.T. for the Naval Architecture program, but they also assigned a Navy Captain and a Navy Commander to M.I.T. as Assistant Professors. These two also supervised the U.S. Navy officers. They were happy to include us Coast Guard officers and take us under their wing. The captain explained how a full load at M.I.T. was considered to be four courses each semester, but he expected everyone to take five courses. He also indicated that the Navy needed specialization in certain areas, such as hydrodynamics, ship structures, ship propulsion, and electrical engineering, plus a few "big picture" officers who could run a Navy ship design and acquisition project. Then he surprised me and said that he would be deciding which officers would concentrate in each area. John Kaplan and I looked at each other, and both wondered the same thing: Does this guy have any actual authority over us?

We had no illusions that M.I.T. would be easy, and I was looking forward to really digging into the class topics and doing things that we never had time for at the Coast Guard Academy, like reading the textbook! At some point over the summer classes, the Navy captain came to our shared office (courtesy of the Navy) to discuss our studies. We very respectfully thanked him for caring about us and including us with the Navy officers, but explained to him that we were responsible to the Naval Engineering program at Coast Guard Headquarters and would only answer to them. We explained that we were taking the Coast Guard option and would not be following any Navy-prescribed course of study.

I half expected him to invite us to find our own office space then, but he took it well. Over our two years of studies, the Navy continued to include us in their activities and camaraderie, but let us follow our own paths academically. That worked well for us. Both John and I decided to attempt to get a Masters in Naval Architecture and a Masters in Mechanical Engineering, concurrently. It took some careful planning, but was achievable. One tricky part was finding a funded research project that applied to both degrees. John's research involved computer modeling of the fluid flow across a submarine hull that was descending at an oblique angle. It was a clear Naval Architecture fit, and he found a professor in Mechanical Engineering who recognized that hydrodynamics and turbulence applied to that discipline as well.

I signed on to an ongoing project in the Mechanical Engineering Department funded by the Department of Energy with the money funneled through a tech firm in Cambridge. The project was to evaluate the efficacy of burning finely ground coal dust in a diesel engine. My specific work looked at the properties of various coal dust mixtures suspended in water as they were injected into a pressurized vessel to simulate the combustion chamber of a diesel engine. I varied the injection pressures and the chamber pressures and compared the different coal mixtures. That whole "trained initiative" thing worked well for me, as there was no guide or prescribed plan for the work. I was the designer, engineer, safety guy, and manual laborer all in one. The project was within the Mechanical Engineering Department, but I found a Naval Architecture professor who accepted that this project could be applied to ships even though coal dust would likely only be used in very large diesel engines if at all practical.

In the process, I learned a few non-academic things at M.I.T. The senior professor who ran my lab was quite happy to have a military officer around, partly because we didn't need to pay me a living stipend that came out of the research funding, but mostly because we were on a timetable and were highly motivated and focused to get the work done. That served as an example for many of the civilian graduate students who seemed content to drag their research out for many years. The divide between us and civilians hadn't been pointed out quite so clearly before.

Additionally, I developed a disdain for some of the Navy and Coast Guard students who would not take a class unless they had "the gouge," a collection of lecture notes, homework solutions, and tests supplied by Navy students who had previously taken the course. I'm sure "the gouge" made it a bit easier, especially if the professors were too lazy to change test problems from semester to semester, but it never felt right to me. I followed my own path and took the classes that I wanted to take regardless of "the gouge."

John and I both graduated with two degrees in the spring of 1986 at an outdoor graduation ceremony in the pouring rain. While there were and still are some seriously scary-smart people at M.I.T., meteorology was apparently not a well-developed academic discipline at that point in time.

FIVE

The Payback Tour (1986-1991)

W e both got orders to Coast Guard Headquarters for our "payback" tour, which was not a big surprise. What was a surprise was that I was ordered into a Naval Engineering billet in the brand-new Office of Acquisition. I had no idea how my studies would relate to Coast Guard acquisitions, and I pushed back with the assignment officer. But he either had no discretion or just wasn't interested in anything that some fresh-faced lieutenant junior grade had to say. Peggy and I looked for a house but settled initially for a small apartment in Silver Spring, Maryland. I reported to Coast Guard Headquarters, and Peggy got her first physical therapy job at Children's National Hospital in D.C. with her newly earned Master's Degree in Physical Therapy.

My job was in the Quality Assurance Division in the Office of Acquisition, which dealt with major Coast Guard purchases: boats, cutters, aircraft systems, software systems, etc., that exceeded a certain dollar amount. One branch of the division reviewed draft contracts to ensure they had clearly described requirements for what was being purchased and appropriate quality system requirements. The other branch conducted quality audits on contractors and potential contractors to verify their compliance with the quality and performance provisions in the contract. In that regard, we worked closely with the contracting officers who bore responsibility for buying what the Coast Guard was ordering. It still didn't seem like a good fit for me, but orders were orders, and I wasn't going to do anything half-assed.

So, I immersed myself in contract rules, learned about quality systems and quality control, and learned how to conduct audits. I got proficient at evaluating what a company's quality assurance manual dictated compared to what was actually happening. It was eye-opening at a time when a lot of U.S. companies were trying to improve quality and reliability to become or stay competitive. Other companies, especially some of the smaller or younger ones, were trying to put on a good fake. Some got by because they just got lucky and had good people working for them. Others had difficulty understanding that you couldn't just inspect the quality into the product at the end of the fabrication or construction process. Still, others tried hard to develop, build, and deliver quality products that were reliable and performed well.

The other junior officers, chief warrant officers, and lone chief petty officer in that office were a good and motivated group. We did have a bit of fun, too, to keep things from getting too mundane. I admit that I had a hand in more than a few pranks. But the most classic prank of all was managed by Pete Oittinen. The division chief, a captain, could be an ass on occasion and was prone to yelling and screaming. To mess with him, Pete quietly set the office clock five to six minutes fast one day, which caused the captain to sprint out of the office thinking he was late for a meeting with the admiral, only to get there are find he was early.

The next day, Pete quietly set the clock five minutes slow, and the captain was actually late for a meeting with the admiral. While he was gone, Pete corrected the clock. The captain came back in a huff, trying to reconcile the clock and his watch. He bombastically demanded that the clock always show the correct time. So, Pete brought the clock home with him, reversed the wires so the clock would run backward, and used a copy machine and scissors to replace the clock face with one having numbers in counterclockwise order. Then he quietly rehung the clock in the office. The captain nearly blew a gasket when he saw it until Pete pointed out to him that the DOD Military Specification, i.e., the "mil-spec," for clocks did not specify rotational direction. The backward-running clock met the mil-spec requirement. Pete dodged a bunch of trouble by turning the clock into a legitimate point about how we should use mil-specs cautiously and not overly rely on them. The captain allowed the clock to stay, although it took a bit of practice to look at it and figure out what time it was. And knowing the backstory, we all chuckled every time we looked at it.

In between a few random pranks, I was working hard to learn everything I could and become proficient at conducting quality audits. I took courses on contracting, courses on quality management, and even a course on software development controls. And I went on every trip I could to gain experience. Sometimes, travel sounds like fun and excitement. Mostly, the traveling part was tedious and exhausting. When Peggy was nine months pregnant with our first child, I did not want to be away overnight, so I booked the trip and flew from Washington, DC to New Orleans before dawn, rented a car, drove two hours to a shipyard in rural Louisiana, conducted an abbreviated quality audit, and returned home just before midnight.

About two years into that four-year tour, I shifted to the other branch. It meant less travel, which was a good thing, but more frustration. I was assigned to work on a few new contracting projects: one to replace the old ocean-going buoy tenders with more modern and more capable vessels; one to build more capable self-righting motor lifeboats for search and rescue work in some extreme seas; and one to build a new Coast Guard icebreaker. Finally, I had some work that was at least somewhat related to ship design! Those projects had me working alongside Coast Guard Naval Engineers again.

In theory, the Coast Guard Operations staff that determined the operational needs and policy would specify the minimum requirements, and then the engineering staff would draft the detailed wording for the technical parts of the contracts. In practice, some of the operations people tried hard to find fancy words to simply say, "Just give us more of what we have already." Others wanted new and improved vessels, but felt the need to specify way too many details that robbed the builder of flexibility in the design. The engineers tended to dust off wording used in previous contracts with only minor modifications. Again, many of the detailed requirements were overly prescriptive, even down to the exact type of hydraulic oil that had to be used. I spent long days in meetings in secluded conference rooms with the contracting officers, operations specialists, and engineers reviewing each of several hundred pages of detailed technical sections. We usually met all day, and I reviewed the sections for the following day in the evenings at home. After several days like this, I started to get the impression that I was the only one who had actually read the sections ahead of time. I thought it was kind of funny and a poor reflection on the process as a whole, but there was no way I was going to let anyone else outwork me or wear me down.

I was a total pain in the ass, dutifully pointing out every time they used vague and unenforceable wording like "the contractor shall build it to best marine practice." I also identified numerous conflicting requirements between sections or within sections. In all fairness, I usually suggested alternative wording, but in some instances, they just didn't know what they wanted, and in frustration, would tell me that the builder would understand and get them what they wanted. The contracting officer would often step in at that point and send them back to the drawing board. The 47-foot self-righting motor lifeboats were built and are still in service, especially in the Pacific Northwest, where the waves breaking over the sand bars at the mouths of the rivers can be treacherous, and boats with this capability are critically important. The buoy tenders were also built and are serving well. It took a 225-foot vessel to replace a 180-foot vessel, but it was past time to replace the old 180-foot buoy tenders, some of which had been built before World War II. The icebreaker was also built and is still in service. I'd like to say that I had a big hand in those projects, but my role was small. It was serious work at the time, but a lot more people had much bigger roles in their success than I did.

One thing that I did take away from the experience was that the Coast Guard's Naval Engineering program was still not very impressive. They were good people who were trying hard, but the leadership within that program at that time just seemed stuck in their ways. They also seemed to be riding the wave of greatness associated with the internal Coast Guard design of the Polar Class icebreaker hulls. It was a novel design and was exceptionally effective at breaking ice, but the design effort had been completed in the late 1960s or early 1970s, and it just seemed like it was a long time for them to rest upon those laurels.

So, as I was approaching the end of my four-year tour, I asked my Naval Engineering assignment officer if I could be considered for a switch to the Coast Guard Marine Safety program for a tour. The Marine Safety program was the other program that sent Coast Guard junior officers to graduate school for Naval Architecture and Marine Engineering. That program inspected commercial vessels for compliance with federal regulations at the design level and the operational level. It seemed like a possible good fit for me. The Naval Engineering assignment officer seemed amenable and confirmed that the two programs exchanged a few officers every year. When I spoke with the assignment officer who handled Marine Safety assignments, he also seemed interested and indicated that the Marine Safety Office in New Orleans could use me.

I was told orders would be forthcoming, but they weren't. I followed up several times and was told that they were just running behind on issuing orders. With no orders in hand and with it getting quite late in the transfer season, I eventually received the dreaded call from the Marine Safety assignment officer, "Can you come to my office so we can talk?" It seemed that the two programs only swapped junior officers on a one-for-one basis, and I was the odd man out. He told me to go talk with the Naval Engineering assignment officer again, where I was informed that most of the billets were already filled, so the best he could do was to transfer me to another Naval Engineering billet at headquarters. I was not happy. It felt like a low point, and for the first and only time, I considered leaving the Coast Guard.

I went back to my office and requested a meeting with my new division chief, a captain who happened to have an aviation background. I told him that I had just discovered one of the reasons why the Naval Engineering program was having trouble retaining their officers: they couldn't manage the people they had and were jerking me around. I asked for his assistance, and he asked for my bottom line. I told him that I wanted a Marine Safety assignment and was not fussy about the location. He stepped up for me and worked a deal wherein I stayed working for him for one more year, then, for the next assignment season, I would work exclusively with the Marine Safety assignment officer and would not even have to speak with the Naval Engineering assignment officer. I took the deal and that Coast Guard option. I was very disillusioned with the Naval Engineering program, but not with the Coast Guard. I was opting for another path within the Coast Guard. And based on my experiences with the Naval Engineering program, I suspected that once I was in a Marine Safety billet, they would lose track of me and forget that I was technically still in the Naval Engineering program. And they did, too.

SIX

Down the Bayou (1991-1995)

The following year, I got orders to Marine Safety Office, Morgan City, located in Morgan City, Louisiana. Peggy and I had never heard of Morgan City and had to look it up in an old road atlas. It was well outside of New Orleans, surrounded by symbols that the legend indicated was swamp. Oh Lord, what were we getting into? We were a couple of Yankees going into the heart of Cajun country and didn't know a soul there. Peggy trusted me to go house hunting by myself. She stayed home with our kids, Conor, who was not quite three years old, and Haley, who was fourteen months old. She figured our options would be limited in such a small town. She had no idea how limited.

Even in 1991, the greater Morgan City area had no realtor multi-listing. Each of the five realtors in town had their own inventory of houses for sale. The first four realtors I met with warned me not to talk with the fifth realtor, as he would try to take advantage of me. That was advice I honored, but none of the four had any houses that would even remotely work for us. Every house for rent was tiny, rundown, in a crappy-looking neighborhood, or all of the above. And the houses for sale were not much different.

As I was reaching the peak of my frustration two days into house-hunting, one of the officers at the unit invited me to join him and his family for dinner. He shared the fact that a full half of the houses for sale in the area were for sale "by owner" and that the realtors probably had avoided those streets while driving me around. After dinner, we took a drive around his neighborhood, and I jotted down the phone numbers for two nice-looking houses that were for sale just down the street from him. It was a decent neighborhood, too. The following day, I set up an appointment with the owners and toured one of the houses. It was plain but solid, well-maintained, and big enough. I conferred with Peggy, and we made an offer that was sealed with a handshake. The local bank had the best rates for a mortgage and conducted the closing at minimal extra cost. Small-town Cajun America was looking okay after all.

After getting quickly settled, I was anointed a trainee marine inspector. Toward the end of my first week, my boss, an old-time marine inspector who had been recalled from retirement to take the Chief of Inspections job in Morgan City that no one else wanted, told me that a ship was scheduled to be inclined that Saturday. He asked me if I had ever seen an inclining experiment. I had not, but told him I at least knew what it was. He said, "Good, because the naval architect coming down from New Orleans to conduct the experiment is not trustworthy and has to be watched closely." So, I went home and cracked open a textbook to review the details of conducting an inclining. It would be my first inclining experiment, and I had to be smart enough about it to make sure it was done correctly. I felt lots of pressure, but I had been in the Marine Safety field for almost one week, and I was already applying my education. It felt like I was in the right field at last. Oh, and the naval architect from New Orleans didn't try to cheat but did try to cut some big corners that I didn't allow.

I was assigned to the offshore section where the inspectors verified that various drill ships, mobile offshore drilling units (MODUs), and oil production platforms complied with all of the applicable federal regulations. I knew nothing about the oil and gas industry, which seemed to be the biggest industry in town, but I was about to learn quickly. We flew offshore in a Coast Guard-leased helicopter that came with its civilian pilot, and I shadowed the other marine inspectors until it started to make sense. The inspections were technical, physical, and detailed. MODUs are loud, busy industrial facilities in a salt water, maritime environment. Most have three tall legs that sit on the bottom, and on which the platform raises to the appropriate height to stay clear of the waves. Each MODU has a massive derrick that supports the drilling operation. We inspected everything onboard related to safety, including the drill floor where the rough-necks worked, the engine rooms, machinery spaces, the accommo-dation spaces, the lifeboat deck, etc. We also crawled into tanks and voids to look at the structural condition of the MODU, and we observed commercial divers doing underwater hull examinations to check the structural integrity of the hull from the outside.

The days were long, hot, dirty, and meaningful. When we weren't offshore, I was either doing the necessary paperwork from the previous inspections or studying the regulations and asking lots of questions. The other inspectors had great experience and a lot of patience to put up with my many questions. I was impressed that their answers usually came with a reference to a regulation or docu-mented Coast Guard policy. There was no shooting from the hip and no guessing about what was required. It took a lot of work and many months, but I was eventually qualified to lead my own inspections.

A newly assigned replacement Chief of Inspections arrived at the unit and implemented a more formal training program complete with qualification boards. The boards were comprised of other inspectors who grilled new inspectors and withheld qualifications if the answers were insufficient. Impressively, several of the questions were judgment questions along the lines of, "What would you do if…" They were meant to test the inspector's ability to think on his feet and apply the rules in more abstract situations. Under the new training program, I earned qualifications to inspect barges, small passenger vessels, and offshore supply vessels (OSVs), and to conduct tank vessel examinations on foreign-flagged tankers. It was a terrific foundation to build on, aided by the fact that we were so busy. With over twenty drydocks in the area and hundreds of commercial vessels to inspect, a trainee could see a complete inspection every morning and another every afternoon for five days a week. We heard that in other ports like New York, inspectors were arguing over who would get to do the only inspection happening that day.

The other inspectors were a mix of junior officers and chief warrant officers (CWOs). The CWOs were guys who had risen through the enlisted ranks and been promoted to CWO. Collectively, they had a varied assortment of Coast Guard backgrounds and came with their own strengths and weaknesses, but they trained and were qualified to the same standards as the rest of us. They were the backbone of the Marine Inspection field because they would serve at field units in inspection billets for the rest of their careers, whereas we junior officers would rise in rank and eventually move to leadership positions that didn't allow as much actual inspection work.

I found the marine inspection work stimulating. We dealt with real issues and real vessels daily. It was challenging work, especially in the Louisiana heat and humidity, but the experience was without compare. One day, I would be conducting a hull inspection on an aluminum crew boat that had been pulled out of the water on a rail system. The next day, I'd inspect a "liftboat," a three-legged industrial support vessel common for shallow water work on oil production platforms that would plant three feet on the ocean floor and then jack itself up on its long vertical legs. They were unique designs with unique problems and risks.

The following day would be inspections on two different offshore supply vessels at their respective docks. The pace was fast, and you had to know which regulations and policy letters applied to which vessel, depending on the vessel type, age, and use. I particularly enjoyed the interaction with the vessel crews. They were mostly locals from Louisiana and Texas just trying to earn a living, and I appreciated their practical way of looking at things. When conducting a fire drill, I asked one crew how they would enter the engine room to fight a fire there. They looked at me and said they weren't paid enough to enter the engine room if it was on fire. Okay. We moved right into the abandoned ship drill then.

On Sunday, August 23rd, 1992, we had just gotten home from a weeklong trip to visit Peggy's parents in Sarasota, Florida. I was out doing yard work when Peggy said that the unit was on the phone. The caller informed me that an elevated hurricane condition had just been set. I had no idea what that meant, but I hoped that the all-hands meeting first thing Monday morning would explain everything. It did. Hurricane Andrew was heading our way, and the entire unit would be evacuated. Arrangements were being made to evacuate us and our families to Fort Polk, located further inland and to the northwest of Morgan City. We prepped the house a bit, grabbed a few things, packed up the kids, and headed to Fort Polk early that evening.

We arrived just after dark and were the first Coast Guard family there. Sure enough, the Army was expecting us. We were directed to some old wooden barracks buildings and informed that women and children would stay in one and men would stay in a different building. They opened the doors and turned on the lights. We watched dozens of large Louisiana-sized Palmetto bugs (giant roaches) scatter. That prompted Peggy to say there was no way I was leaving her and the kids alone in this building overnight. So, we gathered some pillows from a pile in the corner, found a suitable room, and bedded down for a warm, sleepless night.

Early the next morning, other military personnel started to arrive from Navy units in New Orleans as well as some other Coast Guard families. The idea of husbands being separated from their families was understandably still not acceptable to Peggy, so I went and spoke with one of the Army civilians who seemed to have some pull. He confirmed that one of the nearby barracks buildings was empty and available, so I instructed him that it would be the building for all Coast Guard people and families. It made more sense since we all had families with us, and since the Navy personnel were mostly single sailors. I told him we were taking the Coast Guard option. The entire building became the Coast Guard building for the duration.

We made it work despite it being an open-bay barracks left over from World War II, with only one communal bathroom at the back of the building. Families pushed cots together and made the communal bathroom work without anyone getting exposed or being exposed. It was a bit of a unique bonding experience.

For whatever reason, I turned out to be the most senior Coast Guard officer there as a lieutenant. We had no idea initially where the Commanding Officer, Executive Officer, or any of the department heads had gone. We were secure but had a hurricane bearing down, so we waited and wondered how our ancient wooden barracks building would handle the wind.

When the hurricane was scheduled to make landfall and head inland, we shifted to a nearby modern recreational building on base and rode it out without incident. For the next few days, a handful of us junior officers were on the phone trying to track down the other unit members who were scattered all over the state, and we were mostly successful. We only stayed another few days, then headed home once the roads were clear to see what condition our homes were in. Our house lost several shingles. The attic vent fan, the turbine kind that rotated when the wind blew, was missing and likely halfway to Texas. Our carport was down, but the house was dry inside and livable, albeit without electricity. It was still August and hot as blazes, but we were home.

I watched closely and saw that the Coast Guard temporarily split our geographic area of responsibility between the units to either side of us, Marine Safety Office New Orleans to the east and Marine Safety Office Port Arthur to the west. Meanwhile, our unit was stood down for about a week while we got our individual lives back in order. I had never heard of anything like that before and laughed to think the Coast Guard was "taking the Coast Guard option." Good on them!

When we had our houses as watertight and safe as possible, a lot of us pitched in to help other Coast Guard families that needed a hand, as well as some of the civilian employees at the unit. There were hundreds of trees down, and many were up against or lying on top of houses. We all got pretty familiar with using chainsaws over the next few days.

People all over the country were sending donations in the form of food and clothing, and the Red Cross was overwhelmed. We gave them a hand unloading boxes of canned goods and putting them in some sort of order. After a while, the stack of emptied boxes of every shape and size was pretty big. One of the Red Cross workers scratched his head and said he had no idea what to do with all of the boxes. That was easy. I asked him about the dumpster in the parking lot across the street. It appeared to be empty, and with limited electricity in the area, the store that owned it wouldn't be opening or needing it anytime soon. He was uncertain about using it without permission. Not to worry, I told him we were taking the Coast Guard option. We would break down the boxes and put them in the dumpster. If needed, we would ask for forgiveness later. We cleared his box problem and never heard a thing about it.

After a little over a year at the unit, I was moved up to be the Assistant Chief of Inspections, a position that I felt extremely unqualified for. I was replacing a guy with twelve years of marine inspection experience. I had just gotten all my inspection qualifications a few months before. The guy I replaced was being moved back to doing inspections, a clear reduction in his responsibilities. I suspected it was because he was not a stickler for details, always looked disheveled, and paperwork was not his forte. I only learned later that he had a drinking problem and was being placed in an informal probationary status as a sort of last chance.

I did the best I could as the Assistant Chief of Inspections, but felt seriously underqualified. I worked all day and reviewed inspection paperwork and reports each evening at home. I gained a different kind of experience and some insight. It worked largely because the "bullpen" of junior officers and CWOs who did the inspections were well-trained and focused. And there was very little bullshit at the unit.

The inspectors stayed extremely busy with routine vessel inspections, tank vessel examinations, and hull examinations of vessels in one of the numerous drydocks in the region. There were also new construction inspections on barges, small passenger vessels, liftboats, offshore supply vessels referred to simply as "OSVs", casino boats, a molten sulfur ship, a commercial icebreaker, and a massive one-of-a-kind tension leg platform destined for an oil field offshore. The inspectors were regularly swamped, and we tried hard to get the rest of the Coast Guard to acknowledge that and send us another few inspector billets. It never happened.

Instead, the Eighth District Commander, Rear Admiral James Card, came up with a novel idea. What if we identified the best OSV operating companies and allowed them to basically inspect themselves with just a little Coast Guard oversight? That would free up Coast Guard inspectors to focus on the companies that needed more time and inspection. It was an interesting idea. We all knew that a couple of specific companies virtually never had discrepancies on their vessels when we inspected them, and if they did, they were minor and were quickly corrected. Those companies would benefit by not having to take the vessel off the job for a few days to be prepped by the crew and then inspected by the Coast Guard. While off the job, the vessel was not making any money. Here was a chance for them to see less of the Coast Guard and make more money. The only question they had was, "What's the catch?" That was to be determined.

The good news for us at the unit in Morgan City was that we might get a bit of workload relief with this approach. The bad news was that we would have to lead the effort to establish the program, dubbed the "Streamlined Inspection Program," and it would take a buttload of work to get it going.

The Eighth District staff made sure that we had the authority to make it happen and participated in various initial meetings with the

OSV operating companies that we approached. But we did all of the grunt work. The basis was that we would assist the OSV companies by documenting exactly how we inspected an OSV and what to look for. Instead of an annual inspection, the company would be required to implement a more frequent inspection system where certain things would be inspected by their crews at specified weekly, monthly, quarterly, and annual intervals. They would also have to document their inspections. Once established, a Coast Guard inspector would visit the vessel once per year, audit their records, verify that the crew knew how to conduct a randomly selected inspection, and be gone in an hour. But before we got to that point, those same Coast Guard inspectors had to spend a lot of time shepherding and mentoring the companies to give them the best chance of success.

It took months of dedicated work for our inspectors to write down and double-check the step-by-step instructions on how to inspect fire extinguishers, radar systems, fire hoses, ring buoys, fire pumps, engine shutdowns, bilge systems, steering systems, cargo systems, tank vents, crew licenses, crew competency at fire drills, etc. It was an extensive list of things to inspect. And we had to figure out the appropriate frequency for each inspection. Then we had to let some folks from the OSV companies review them to make sure that we didn't inadvertently use too much Coast Guard jargon or slang.

When we were finally ready to launch the initiative, we let each company put one vessel into the Streamlined Inspection Program. The plan was to move slowly and carefully. That worked until about three months later. At a regularly scheduled meeting of our working group of Coast Guard and industry personnel, one of the companies asked if they could add another vessel or two unofficially. I hesitantly asked, "Why?" It seemed that their crews really liked the program. Comments from crewmembers indicated they felt tremendously empowered by the Streamlined Inspection Program and loved it. The companies reported that all of the equipment and systems were verifiably in compliance with the regulations, plus the empowered crews were taking ownership of their vessels more than ever. Also, the overall condition of the vessels was improving across the board. One of the companies reported that the program just saved them thousands of dollars the week before, when the crew reported that the anchor windlass sounded extra loud when they conducted a required inspection as part of the Streamlined Inspection Program. Sure enough, there had been a problem with the gears that was easy to correct but would not have been identified or fixed otherwise until the anchor windlass completely failed. Talk about unintended consequences!

So, we skipped the remaining trial period and made the Streamlined Inspection Program official. Last I checked, it was still in place and had spread to numerous other types of Coast Guard-inspected vessels across the country.

Somewhere in there, I was promoted to lieutenant commander. It was a big promotion personally because it meant that I could stay in for 20 years and retire with a pension. It meant financial stability down the road, even if I was never promoted again.

As a lieutenant commander, I would be in a leadership position. I had not been to any of the formal Coast Guard leadership courses beyond what had been taught at the Academy. Frankly, I wasn't that impressed based on what I heard from people who had attended any of those one or two-week leadership training courses. The course development folks and the instructors no doubt did their best and tried hard, but the results were mediocre enough that I didn't want to invest the time. But I knew I needed to improve my leadership skills, so I took it upon myself to get better by reading a lot of books. I avoided books on leadership and management and avoided fiction almost entirely. Instead, I read biographies, autobiographies, military history, and more military history. I was a military officer, so I needed to read something relatable, not a biography of the CEO of some large company. I read what Theodore Roosevelt wrote, what Omar Bradley wrote, and what Joshua Lawrence Chamberlain wrote about his Civil War experiences. I read about Jonathan Wainwright in World War II and about William Wallace, Okinawa, and the Battle of the Bulge. I read what Bruce Catton had to say about the Civil War and the generals, and what Barbara Tuchman wrote about the Founding Fathers. I studied what Samuel Eliot Morison wrote about the Navy in World War II and what William Manchester wrote about MacArthur. Lessons were there to be learned. All I had to do was pore through them, and I did so intently. It was the start of a life-long study that continues to this day.

Late the following spring of 1993, I got a call from the Commanding Officer to immediately come down to his office. Nothing braces you up like that kind of call. It sounded serious, but I knew I hadn't done anything wrong. On the way down, I passed my boss, the Chief of Inspections, who wanted to talk about something. I put him off because the C.O. had said "immediately." The C.O. and X.O. were waiting for me in the C.O.'s office. The C.O. calmly told me that he had relieved my boss and that I was the new Chief of Inspections. He shared nothing else about why. There were no details to follow. He also added that he knew I was still relatively inexperienced, but could come to him with any questions, technical or otherwise. All I could say was, "Yes, Sir." My now-former boss was immediately transferred to the Eighth District staff in New Orleans, and I only ever saw him one other time. I was a brand new lieutenant commander in a commander's billet and grossly unqualified. It was scary but also somewhat exhilarating. The C.O. had faith in me and trusted my judgment. I took a deep breath and moved forward.

The first thing I had to do was inform the other inspectors, and then find a new Assistant Chief of Inspections. The other inspectors took it well enough. We were all extremely perplexed over the firing of the previous guy and were all equally in the dark. He had been well-liked, knowledgeable, and even-handed. He had installed a great training program that worked well and produced results. He brought the entire department in line with how the rest of the Coast Guard conducted and documented marine inspections. It really started to hit me that I had big shoes to fill. All I could do was just roll up my sleeves and get to work.

One day, I got a call from one of the inspectors who was on a foreign-flagged tanker offshore and finishing an inspection. As he was flying off the tanker in the chartered helicopter, he saw an oil sheen in the water near the stern of the ship. He directed the helicopter to return to the tanker so he could investigate. He suspected there may have been a stern tube leak. The stern tube was the tube through which the propeller shaft passed on its way from the main engine through the hull to the ship's propeller. The tube was filled with an oil for lubrication purposes and sealed at each end. On occasion, the outer seal would develop a leak, and some of the oil would leak out. But on detailed examination, the volume of oil in the stern tube on this tanker had not changed in weeks, and the crew had not added any to the amount. That's when he called me. There was a bigger problem because the oil could only be from either a leaking fuel oil tank or a leaking crude oil cargo tank. That implied there was a hull breach of some kind.

We required the ship to hire a commercial diver to inspect the hull from the outside. He found the leak. It was from a centerline tank midships. The bottom of the cargo tank was the actual hull, as this was before the requirements for double-hulled tankers were applicable. There were several large corroded and pitted areas, and one of those areas was the source of the leak. In plain words, the bottom of the ship had corroded completely through. And the corrosion was at a location where hull stresses during loading and unloading operations could be highest. In bad weather during voyages, the hull stresses could also spike. That meant the weakened hull could very quickly develop a crack, which could lead to a major failure of the hull and a massive oil spill.

At our insistence, the ship's owner provided a copy of the most recent drydock report. Surprisingly, the ship had been in a drydock for repairs immediately before it loaded this cargo. The report also indicated that some hull repair work had been done in this area. We were disappointed that the classification society surveyor had insisted on only limited repairs and had not required a more extensive assessment of the hull thickness.

I recommended to the Commanding Officer that we throw the ship out of port and out of U.S. waters and force them to take their very marginal ship elsewhere. He listened patiently while I made my case. It was a technically sound recommendation that was clearly within his authority. But he saw the bigger picture that I didn't have the experience to recognize. He pointed out that if we sent it somewhere else and the ship ran into rough seas on the way, the ship might break apart, and the crude oil would spill. He felt the safest and most prudent course of action would be to allow the ship to offload its cargo as planned, as the weather there off of Morgan City was excellent. Then he would throw it out and restrict it from returning to the U.S. until a proper hull assessment was done and full repairs made. It made sense and was a valuable lesson for me about the judicious exercise of authority, and the Coast Guard option.

The tanker offloaded without incident. The owners then took it out of service and immediately scrapped it. The second lesson from this case left me shaking my head. These owners knew the ship had major issues and had simply tried to squeeze one more very profitable trip out of the old hull. For them, it was about making money regardless of the environmental risk.

Heading home from the office one day as I drove over the bridge across the Atchafalaya River in downtown Morgan City, I saw an offshore supply vessel that appeared to be overloaded. These OSVs all had superstructures forward and open, flat decks aft to carry cargo for MODUs and oil production platforms offshore. This one particular vessel appeared to be very low at the stern. OSVs all had a required white horizontal line painted on the stern indicating the maximum that the vessel could be loaded, i.e., the white line had to be above the water and not submerged from the weight of the cargo. That would have been a violation of the vessel's Stability Letter issued by the Coast Guard to each vessel indicating its maximum loading conditions. For OSVs, the stability letters had a standard format, but the details varied greatly by vessel according to the vessel design, its liquid cargo capacity in tanks below the main deck, and its capacity for solid cargo stacked on the main deck. More important than whether it was a violation of the Stability Letter was the fact that it jeopardized the safety of the vessel and crew. I suspected the problem was a lot bigger than just this one vessel.

A few of us came up with a plan to see how widespread the problem was, to better understand the problem, and to enforce the provisions in the vessel Stability Letters. We would take the Coast Guard option and use our authority to do something totally unique that might shake things up. The plan was to set a date on which we would not schedule any vessel inspections. Instead, I would send all of our marine inspectors out to the vessel loading ports to go onboard OSVs as they were loading. We would verify that the Stability Letter was onboard as required and ask the vessel's captain to demonstrate that the cargo loaded or scheduled to be loaded was within the parameters specified in the vessel's Stability Letter. We would not ask permission from the Eighth District unless we had to. They would only complicate things and would add no value if they even agreed.

It was a risky plan because we could seriously stir up some shit with the industry that was already quick to complain. While the Coast Guard regulated the commercial shipping industry, we also tried to work with the industry to ensure compliance and to understand their challenges. As a result, Coast Guard senior officers regularly attended meetings with industry representatives to discuss new policies, address industry concerns, and be available to hear them out. Those meetings would provide a handy forum for the industry to complain if our plan shut down too many vessels or caused major delays in getting cargo offshore. The OSV companies were all about meeting the needs of their customers. Those were the oil companies that paid big money to the MODUs to drill for oil. The MODUs, in turn, needed a steady supply of consumables like fuel, liquid drilling mud, and drill pipe, and they tolerated no delays from the OSV companies trying to keep them supplied. Time was money, and not much stood in the way of the drilling. If our plan resulted in delays to critical supplies being delivered to a MODU just because of a "paperwork issue" or the vessel captain's incomplete understanding of the Stability Letter requirements, or a Coast Guard marine inspector's over-zealous interpretation of something, the cost from the delay could be thousands of dollars per hour. And my boss would get an angry phone call.

So, I briefed my Commanding Officer on the plan and answered his questions. I asked for his permission to proceed, which was necessary because if he agreed, he was essentially giving me enough rope to hang both of us. There were several ways this could go sideways, and each way ended with both of us getting lambasted by the industry and then by the rear admiral at the Eighth District Headquarters in New Orleans. After ensuring my C.O. that we would not look for reasons to tie up commercial vessels and that I had controls in place to ensure an inspector wouldn't get carried away, he granted me permission for the operation as intended.

We planned the operation about six weeks out to give me plenty of time to distribute the plan and explain it to all of the inspectors. It was new to them, too. We even gave the operation a name. This was too delicate an operation to trust to a written plan alone, so we had a couple of sessions where the concept was explained, questions were answered, and potential problems were identified. The more experienced CWOs were skeptical that the effort would produce any meaningful results, but were willing to give it a try.

On the appointed day, I sent the inspectors all over the region to every loading port and oil company terminal. I sat nervously near the phone, waiting for the first call that there was a big problem.

We rarely saw OSVs at the oil company terminals. Instead, we usually inspected them at either a shipyard or at the OSV company dock when they were unloaded. By going to the oil company terminals, we would see them being loaded or already loaded and about ready to get underway. We learned a lot. First off, most of the oil company personnel did not keep track of the carrying capacities of the various OSVs. All cargo heading to a certain destination was stacked together either in an adjacent building or, more frequently, inside a large, spray-painted square on the concrete dock. When an OSV pulled up that was chartered for the destination, everything in the square was loaded onboard. The captains were reluctant to say "no" to any cargo because they didn't want their OSV to get fired from the job. Further, some of the captains had difficulty determining whether the cargo exceeded the limits or not. Understandably, it could be tricky to determine the vertical center of gravity for the combination of stacks of drill pipe, large bags of cement, pallets of dry material, and sometimes a new subsystem consisting of a vertical pressure vessel with associated piping and structure. We also learned that the more remote the oil company terminal, the more likely that the capacity of the OSV would not be checked relative to the cargo intended to be loaded.

Only one vessel was missing its Stability Letter, but the captain was able to obtain a copy of it from his company office pretty quickly. One vessel captain laughed when asked to prove the deck cargo was within the limits. He referenced his lack of a high school diploma and his general inability to do math. But he said he called the engineer in the office and had that guy figure it out for him whenever there was a question. That worked, and the load was proven to be within the Stability Letter limits.

Overall, it was a great introduction to the human factors component as it relates to commercial shipping. All of the technical requirements matter little if the mariners and their employers didn't, wouldn't, or couldn't comply with them. Plus, it brought into focus the pressures faced by mariners and by companies on occasion. For the mariner to say "no" to carrying a specific cargo was to risk being fired. If the company backed their Captain, there was a risk that the company would be fired from the job. That infrequently happened, but it was still a legitimate risk that they could not dismiss lightly.

Life in Morgan City had some unique challenges. Amenities were limited. Restaurants were limited. Some only offered a variety of seafood, but only if you wanted it fried. Peggy regularly had to go to more than one grocery store to get everything on her normal shopping list because no one store carried it all. In the "summer" half of the year, the temperature and humidity would frequently match at 95. There was an initial lack of friendliness toward people who weren't from there because of the recent oilfield boom/bust swings that alternately brought in outsiders who tried to take advantage of Cajun hospitality and who then departed during the bust, often owing money to locals. It took them a while to warm up to us Coast Guard outsiders. And the kids started picking up a little Cajun accent with certain words!

At some point, I began referring to Morgan City as "beautiful Morgan City" when I spoke with other Coast Guard units on the phone. It was mostly in jest because Morgan City was generally seen as a uniquely undesirable location to be assigned. The town was slightly quaint but not very picturesque, and the big annual event was the Shrimp and Petroleum Festival, which delivered some good food but was otherwise not very impressive. But I also said "Beautiful Morgan City" because it always got a laugh, which I hoped made my call harder to forget. When leaving a message with a secretary (they had such essential people back then), I jokingly insisted they write on the phone message that I was calling from "beautiful Morgan City," and I swear that a lot of people returned my calls who otherwise wouldn't have because of that. I'm sticking with that story, too.

As the end of my 3-year tour approached, I started a dialogue with the assignment officer. It was his job to find a full commander to assign to the billet I was filling. It was proving to be quite a challenge for him. It was difficult enough to find volunteers to transfer to Morgan City. It was a small town off the beaten path and not seen as a stepping stone assignment. Plus, by the time most officers had made commander, they had served 20 years and could retire in lieu of accepting orders. Short of answers, the assignment officer asked if I would consider staying there one more year. He promised that if I did, he would take care of me the following year in terms of assignments. I discussed it with Peggy. On one hand, we were ready to leave, but on the other hand, if we stayed, I would have been a

Department Head for two years and could legitimately ask for an assignment the following year as an Executive Officer at a small Marine Safety Office. We decided to stay. It seemed like a good career move and, truth be told, I liked the job in Morgan City. The challenge was that it meant our son, Conor, would go to first grade in a school not known for its high standards and in a state well known for its low academic standards.

During that fourth year, my Assistant Chief of Inspections, a lieutenant, came to me with a problem. He was a great officer and had tremendous experience that I leaned on. He had recently transferred in from a small detachment in Panama City, Florida, and had left his ex-wife and child behind. The custody arrangement was complicated, and his ex-wife did not know that he had been transferred out of Florida. He drove back to Panama City every weekend to maintain a residence and presence there, but it was getting to be unmanageable. He told me that he would be submitting his retirement request. He had been enlisted before he became an officer and just had 20 years of service, the minimum for retirement. I hated to lose the guy. He was smart and experienced, and I had learned a lot from him in the short time he had been there.

There was no obvious option for me to keep the guy. It was his right to retire, and while the Coast Guard could technically deny his retirement request, that was rarely done and then only for a very good reason. Then I had an idea. Why couldn't I try to get him transferred to Mobile, Alabama? That way, he could stay in the Coast Guard, and his family issue would be substantially easier to deal with. So, I took the Coast Guard option. I floated the idea past the Commanding Officer. He recognized immediately that we would lose the guy regardless, but that the Coast Guard didn't need to lose the guy. He gave me a green light to see if I could make it happen.

The assignment officer was easier to convince than I had expected. Maybe it was because he still owed me for agreeing to stay in "beautiful Morgan City" for another year. Maybe it was just out of a sense of decency. I knew the unit in Mobile would be thrilled to have him, especially as an over-billet, a person beyond their assigned number of people. It took a few weeks, but the assignment officer worked the system and worked on his boss and got approval. We lost a valuable officer, but he stayed on active duty for several years after at the unit in Mobile and then took a civilian position there to train new inspectors and provide continuity at the unit. It was a win for him and the Coast Guard, but a loss for Marine Safety Office Morgan City. It was the right thing to do and the right way to take care of our people, even if it was a unique request and a headscratcher at first.

The following year, the assignment officer was good to his word, and I got my desired assignment as Executive Officer at Marine Safety Office Cleveland on the "north coast." The detailer had finally found a commander to fill my billet in Morgan City, albeit with the promise that he could fleet up into the Executive Officer position in Morgan City a year later. Morgan City was still holding strong as a less desirable location to transfer to.

As a departing gift, one of the junior officers snuck a license plate onto the front of my car that read "Eat More Possum." It was a fitting and funny way to remember Morgan City, and I still have that license plate all these years later. But it's hanging in the garage, not on a car.

The Number Two (1995-1998)

We arrived in Cleveland in the middle of a heat wave. The locals were suffering, but it was nothing compared to the summer in south Louisiana we had just left. Cleveland was, in fact, about ten degrees cooler and significantly less humid than life on the bayou.

Marine Safety Office Cleveland was a small unit of seventeen people. It was one of several similarly sized units spread across the Great Lakes to accommodate the commercial shipping interests and associated workload. My boss was Commander Jack Davin, a genuinely nice guy and a good boss. In his position, he bore several key titles, including "Captain of the Port" and "Officer In Charge, Marine Inspections," as well as Commanding Officer of the unit. Those first two titles were based on laws that granted significant legal authority to Coast Guard officers so designated. That's why the Commanding Officers were commanders even though the unit was small. The Coast Guard wanted the people wielding those responsibilities to have been around the block a few times and have some legitimate experience and proven judgment. It's a unique position because even though he reported to the District Commander, an admiral who happened to be located just a few blocks away in a large multi-agency federal building down-town, the authority was given to my boss. As Executive Officer, or "XO," I was designated as the Alternate Captain of the Port and Acting Officer In Charge, Marine Inspection, in my boss's absence.

The inspection workload, casualty investigation workload, and pollution response workload were notably less than I was used to in Morgan City, but I was kept busy with the administrative and budget hassles as well as serving as a technical consultant on every-thing. I quickly discovered that even though I only had four years in the Marine Safety field, I had seen more inspections and handled more issues than people with many more years of experience. It confirmed that Morgan City was an outstanding place to learn, even if it was a challenging place to live.

Cleveland was a destination port for iron ore coming out of the mines in Minnesota and points further west. It was processed into taconite pellets and loaded by conveyor belt into ships that had been purpose-built for the Great Lakes. These ships were known as "lak-ers." They were big but blunt-nosed to maximize the carrying capacity while still being able to fit into the locks at Sault Ste. Marie located on Michigan's Upper Peninsula. Because the Great Lakes are fresh water, which is much less corrosive than salt water, the fleet of U.S. and Canadian ships that worked on the Great Lakes was a mix of a few newer ships and several fairly old ships. When arriving at Cleveland, some lakers would offload at a terminal on the lake, but others would transit up the Cuyahoga River and offload directly at one of the steel mills still in operation.

The Cuyahoga River lived up to its translation as "crooked river." It was a difficult transit that required ship handling expertise and a good bow thruster, a propulsion unit near the bow of the ship that pushed water sideways to help push the bow sideways. At least one of the bends was so tight that ships had to lean against the steel bulkhead that made up the riverbank on the inside of the turn and pivot around. For the lakers, it was one-way traffic only, and they coordinated transits among themselves.

Near the mouth of the river and right where a smaller waterway, the Old River, merged with the Cuyahoga, was an area called "The Flats" that was a growing entertainment area with restaurants and several nightclubs. Some of them were accessible from the water and had even installed docks so boaters could tie up and spend their money at the various establishments. The problem was that when the weather was good (about three months of the year from my perspective, but about six months of the year for the hardy Clevelanders), the number of moored boats extended so far into the river that the lakers could not pass. The boats routinely "rafted" by tying up to the outboard side of a boat that was tied up to the dock. Having five or six boats rafted outboard of each boat that was moored to the dock created one large mass of boats and stuck out well into the river.

The laker companies were screaming that commerce was being impeded and the boats and boaters were at risk of getting hurt. And that was assuming that the boaters were sober. On the flip side, the economy in Cleveland was starting to turn around, and the city benefited from the proliferation of restaurants and entertainment establishments. The Drew Carey Show was brand new and was set in Cleveland, and the opening for the show was filmed in the Flats, complete with a catchy song and choreography. The Flats was the place to go! The restaurants wanted to know why the lakers couldn't schedule their arrivals for times and days when they weren't busy, or just wait somewhere until the boaters got in their boats and went home.

On top of that, several of the laker captains had also complained that a couple of specific spotlights on top of some of the buildings in the Flats were right at their eye level as they transited at night and were somewhat blinding. They asked us to fix that, too.

Jack Davin and I put our heads together and explored our options. We concluded that it was time to be a little ballsy and take the Coast Guard option. We would use some of our Coast Guard authority while threatening to use a lot more of it if everyone involved didn't come to an agreement. We knew that it would be impossible for the ships to time their arrivals for when boaters were not at the Flats. There were just too many variables: the weather enroute, the timing of getting through the locks at Sault Ste. Marie, the weather in Cleveland that would determine if boaters would be out, the speed of loading cargo at the loading port, etc. And we understood that delays would be expensive for both the ship owner and the steel mills that needed to build up a sufficient supply of taconite by late fall to last them through the winter when the lakes were frozen, and ships couldn't move.

So, we called a meeting of the laker companies and the restaurants in the Flats that had docks. The official goal was to reach an agreement between all sides. But in reality, we would be dictating the solution to them. The plan forced the restaurants to have a designated "dockmaster" who would be responsible for any boat that moored there or tied up to another boat that was moored there. The dockmaster would be employed by the restaurant and would have to listen on a VHF radio for broadcasts from the lakers who made safety calls before the start of their trip up the river. If a laker was approaching, the dockmaster would be responsible for moving boats or directing the boat owners to move their boats out of the way. What made it workable was that the biggest restaurant with the most dock space was located right where the Old River fed into the main part of the Cuyahoga. If the laker was heading up the Old River, the boats could shift to the Cuyahoga side. If the laker was heading up the Cuyahoga, the boats could shift to the Old River side. If both sides were already full, they would have to have some boats get underway and out of the way of the laker.

The restaurants were not thrilled with the idea as it put more responsibility on them, but they liked the alternative even less. We threatened to exercise Captain of the Port authority and, for safety reasons, prohibit all rafting of boats at those docks. That would limit the boats to a very small number that would be one deep instead of five or six deep. And that would mean fewer paying customers. The restaurants agreed. Could we have successfully prohibited them from rafting boats? Possibly. It's always problematic if you block a navigable waterway. We pretty clearly had the authority, but didn't know what would happen if they made a lot of noise, complained loudly to the media, and complained loudly to the admiral. But we stuck to our position and forged an agreement. To their credit, they stuck to it and made it work.

Not long after, we got a report that a high school crew team had been practicing on the river and had been swamped by a passing laker. We knew that the lakers transited at a crawl, so it wasn't as if the lakers had swamped the rowers with a high-speed wake. But we also knew that the shoes worn by crew teams were usually secured into the crew shell, which meant that if the crew shell was swamped, there was a real risk that the high schoolers would not be able to quickly untie the shoes and extricate their feet before getting submerged or worse. That was not a good scenario to contemplate. None of us at the unit had ever rowed crew, but several of us were Coast Guard Academy graduates and, as such, had classmates, friends, and roommates who had rowed crew. We knew their zeal for the sport and dedication to practicing in all weather. The head crew coach when I was at the Academy was referred to as the "Ayatollah" because of the way the crew team followed him religiously. We strongly suspected that the high school crew teams in Cleveland would approach their sport with similar enthusiasm and dedication.

The real problem was that there were only a few places on the river that were safe for a crew shell to wait while being passed by a laker, where they would not risk getting swamped by the water from the laker's bow thruster. So, the first thing we did was sit down with representatives of the ship operating companies. With them in the room lending their expertise, it was an easy thing to identify the safe locations on the river for a crew shell to wait while a laker passed by. The lakers recognized that they did not own the river for their sole use, and they very certainly did not want to hurt anyone. They also suggested that the crew coach in the chase boat carry a handheld VHF radio so he could hear the broadcasts when the lakers were getting ready to enter the river or leave the offloading port upriver. The radio would also give the crew coach the means to communicate with the laker if necessary. Sounded like an easy and good solution.

I called the crew coach the following day. He was less enthusiastic about the plan. I think it sounded like a hassle to him, and he didn't want to spend the money to buy a VHF radio. I explained the situation and how the lakers were restricted in their ability to maneuver in many parts of the river. I told him that we viewed this as a safety issue and that his high school team was at risk of damaging their crew shells, getting injured, or worse. Inexplicably, he wasn't buying it. So, I changed my approach and exercised the Coast Guard option. I told him that if he didn't get onboard with the plan, I would be at the next school board meeting to personally give the school board the official Coast Guard position that the crew coach was knowingly putting his high schoolers in a dangerous position by refusing to take simple and prudent measures that would allow his team to avoid problems from the large lakers that the crew shells would always lose. I asked him if he really wanted to argue in front of the school board with the Coast Guard over an issue of safety on the water. He changed his mind and bought into the program. He bought a radio and used the maps we provided showing the safe locations to wait while a laker was passing. And that was the last we heard of that waterway user conflict.

It was a good tour in Cleveland. I liked being the Executive Officer at a small unit where the administrative workload wasn't overly large, and I could still have a hand in the operations.

And one of those operations involved removing a large amount of oil from a barge that sank in Lake Erie in early 1942. The year before I arrived, the old barge had started to leak, and drops of oil were seen floating on the surface of the lake by a passing helicopter. The unit had done a lot of research, including a trip to the Cleveland public library to dig through newspaper records. The barge sank during a storm, as did the tug that was pulling it, with the loss of all hands. Some oil washed up, but an oil spill in Lake Erie was not a big issue in 1942. The wreck was soon forgotten until the early 1960s when the St. Lawrence Seaway was dredged and then opened to ocean-going ships. Those ocean-going ships, or "salties," had a draft deeper than the lakers that operated exclusively on the Great Lakes, and one of those deep draft ships found the uncharted wreck when they collided with it. Some more oil spilled, but the news at the time seemed more focused on removing the wreck. The U.S. Army Corps of Engineers got involved, and the wreck was refloated but rolled over and sank again. They refloated it a second time, and it rolled over and sank a second time. It became obvious that getting the wreck into a port was not going to work, so the best the Corps of Engineers could do was drag it away from the shipping lanes and let it sink again. There the barge lay until 1994 when the oil drops showed up.

When I arrived in 1995, a plan was in place to use a salvage company and divers to remove all of the oil from the barge before it deteriorated further and spilled all of its remaining contents. The barge was sitting on the bottom upside down, so access to the cargo tanks would have to be through the hull. The salvage company, Donjon Marine, did an outstanding job attaching valves and drilling holes through the hull using a process called "hot tapping." They hot tapped two holes in each cargo tank, sucking the oil out through one while allowing water to backfill through the other hole. When the oil flow being sucked out started to become mostly water, they switched the suction to the other hole and repeated the process until they were confident all of the oil was recovered from the tank. Then, they moved to the next cargo tank. We had a good stretch of weather for the operation several miles offshore. The good weather helped, but the salvage guys knew what they were doing and did it expertly. Not a single drop of oil was spilled during the recovery operation, and more than twice the estimated amount of oil was recovered. The funny thing was that this "waste oil" was taken to a waste oil processing facility in the Detroit area, where it was tested and determined to be as good as new heavy fuel oil currently being produced. The operation was a complete success due largely to the salvage company and the planning that the unit had done before I arrived. Still, the operation took longer than planned and involved a lot of coordination and oversight. Had there been an oil spill associated with the oil recovery operation, it would have been our problem to deal with, and there was a legitimate chance a big oil spill could happen with a thin hull that was already showing signs of deterioration and at a time when public awareness of oil spills was significantly higher than in the 1940s and 1960s. We kept the media and local officials apprised of what was happening and the risks. When the entire operation was over, we held a final press conference for my boss to summarize the operation. We even handed out some small samples of the recovered oil in sealed glass jars just to show everyone how thick and black it was.

After two years, Commander Davin was tour complete and was relieved by Commander Ron Branch. Ron and I had never met, but we immediately hit it off. He was from Oklahoma and had entered the Coast Guard after college through Officer Candidate School. He was a seasoned Marine Safety officer with a real common sense approach and a practical outlook. My third year there was relatively uneventful but enjoyable, and I learned a lot from Ron. We thought alike and had complementary approaches to problems. He had a tenet that fit right in with how I liked to handle operational issues, but I had never heard it expressed so clearly. He said that when the call comes reporting an issue with a commercial vessel or a possible spill of some kind, "We go. We may not learn anything when we get there, but we sure as hell won't learn anything if we don't go." That was perfect. I filed that away and quoted it many times over the subsequent years.

As my tour came to an end, I knew that after two field units in a row, I was destined for a damned staff job. Plus, I had recently been selected for promotion to Commander. There was no way for me to stay operational. My orders were to report to Coast Guard Headquarters to the Office of Design and Engineering Standards. At least I was still assigned within the Marine Safety field, living up to my prediction that the Naval Engineering program would forget about me. I would head the Human Element & Ship Design Division, and I had no idea about what my responsibilities would be.

EIGHT

Back at Headquarters and the DC Grind (1998-2002)

We found a house to buy in Fairfax, VA, and moved the family. It was a fairly straightforward move, and we made the timing work out, but we were not quite ready for the faster pace and higher levels of competitiveness we found living in the greater Washington, DC area. Everything, including traffic, work, who you were meeting with, how good your ten-year-old was at sports, how many activities your kids were involved in, etc., was a competition. It was amusing and exhausting at the same time. It did not lend itself to making close friends, just acquaintances.

The exception seemed to be within Coast Guard Headquarters, located south of the Capitol at Buzzards Point. Everyone seemed focused and serious, but generally approached problems cooperatively and with a true desire to get the job done for all the right reasons. Maybe it was the nature of work in the Marine Safety field without the need for large capital outlays, or the regulatory and policy work that had clear and direct impacts in the real world. We weren't in competition with aviators or the cutter program, trying hard to justify big funding for new equipment and systems. We also regularly met with members of the maritime industry at national organization meetings and official advisory group meetings, so we were not insulated and detached from the effects of what we did.

I was in charge of the "Prevention Through People" program (PTP). It was a unique approach that had been started a few years previously to focus on the human factors aspects of commercial Marine Safety as a complement to the black-and-white regulations that focused heavily on physical requirements and equipment. PTP was intended to prevent injuries and marine casualties through non-traditional means. We worked with several different maritime focus groups in partnership. It was a different approach, but I was intrigued from the start.

I found out pretty quickly that I was picked for the job by the two-star admiral in charge of Marine Safety, Rear Admiral James Card. He had been the Eighth District Commander in New Orleans when I was stationed in beautiful Morgan City. I believe that the human factors approach to improving Marine Safety was a fledgling notion of his at the time, and that he recognized that my efforts in Morgan City to check stability letter compliance fit well into that concept. He was the same admiral who had initiated and approved our Streamlined Inspection Program. Rear Admiral Jim Card was a true Marine Safety expert and a bit of a visionary. He was a good man, too, and I was honored a few years later to be part of his retirement ceremony.

The PTP approach was based on five principles: honor the mariner; take a quality approach; seek non-regulatory solutions; share commitment; and manage risk. The whole approach was vastly different from the basic inspection work where we inspected their vessels for compliance with the regulations, and they had to comply. But the more I read and the more the staff and I discussed it, the more parts seemed not so different. Aspects harkened back to some of my early inspection qualification boards, where I was asked several judgment questions. They weren't questions about when to pull the papers and tie up a vessel. They tended to be related to legitimate workarounds, such as, "If the vessel is required to have thirty lifejackets, but four of them are in poor condition and not usable, what are your options?" Of course, you could say that the vessel was not in compliance with its Certificate of Inspection that stated a minimum of thirty lifejackets were required, so the vessel should be tied up until they replace the four bad ones. But, to quote one of the seasoned CWOs in Morgan City, that would be the "wrong answer." The correct answer was to talk with the vessel captain and company representative, see what kind of job they were on and how many people they would be carrying, and then temporarily limit the number of people allowed to match the number of lifejackets available. There was no need to hinder commerce if there was no safety concern and the level of risk didn't increase. That sort of sounded a bit like a PTP thing. Additionally, the Streamlined Inspection Program was based on the mariners being able to verify compliance with the regulations and being duly empowered. That sounded like "honor the mariner" to me as well as "share the commitment."

A big part of the PTP program was to take a risk-based approach to identifying and solving problems. Identify the risks, assess the risks, manage the risks, and mitigate the risks. We learned about lots of different risk analysis techniques and how they could be used in the Marine Safety field and by other programs within the Coast Guard. We also found a few that were best used by PhDs and only under-standable by PhDs. We had to keep things simple for our own sake as well as that of the users: maritime companies, Coast Guard field offices, individual mariners, etc. They were not too thick to under-stand. They were just too busy to study and train on complex risk approaches when simpler techniques were usable and oftentimes more repeatable. Little did I know how big these risk analysis tools would become in a few short years.

In the meantime, we met with an association representing the maritime towing industry, another association representing the passenger vessel industry (smaller vessels like ferries, dinner cruise boats, and tour boats), and another representing the cruise lines, and we worked cooperatively on different approaches to improve safety.

My new boss was Captain Mark Van Haverbeke, a consummate professional. My staff was good but destined to get better. I took advantage of the opportunity to hire an old friend, Joe Myers, who brought a wealth of knowledge and common sense. I had called him to ask about some professional organizations where we could advertise an open position. He was very interested in the job himself, applied for it, and was substantially better qualified than all the other candidates. Lieutenant Commander Duane Boniface was a brilliant guy who was as comfortable talking with PhDs as he was with industry representatives. Lieutenant Commander Paul Szwed was an old acquaintance from MSO Morgan City and had a ton of Marine Safety experience. He handled issues and questions related to novel commercial vessel designs. I initially passed on the chance to bring Mayte Medina on the staff, but quickly wised up and hired her. She was exceptionally experienced related to the International Maritime Organization (IMO) and how it works, and she became an invaluable part of the staff. On top of that, we had a few junior officers right out of graduate school and two civilian contractors who were technical writers. The technical writers were great and very much needed, as a lot of what we did in the PTP program was about outreach. They could take our concepts and ideas and work them into publications, pamphlets, messages, etc., well beyond what we could do. I had them regularly review my draft correspondence to check that the tone was appropriate. I also had a bit of fun challenging them to work some obscure quotes into various safety articles.

Mayte had her work cut out for her getting me prepared for my first IMO meeting. The IMO is a United Nations organization that deals with maritime issues. It sets international standards through treaties. Like everything else that involves numerous countries and no small amount of bureaucracy, things can move slowly at IMO, and progress can be stifled in a bunch of different ways. I was going as part of the U.S. delegation, would be participating at IMO's Maritime Safety Committee meeting, and would be the U.S. lead in a Human Factors Subcommittee. I was feeling the pressure.

Mayte had to explain the workings and terminology of how IMO meetings would flow and how the morning and afternoon coffee breaks were where consensus was established, and blocs were formed. The initial project involved creating guidance to improve safety based on human factors and targeted to specific audiences. Guidelines were developed for lower-level mariners, senior officers on the vessels, and company management, considering the different perspectives and responsibilities. The learning curve was steep because of how IMO conducted business, but Mayte did a great job ensuring I was as prepared as possible. The usual approach by the delegation was to set goals that were easy to attain and then claim victory. I wanted more than that. I wanted real progress and with Mayte's insight, it was a good meeting.

At a subsequent IMO Maritime Safety Committee meeting, we submitted a U.S. position paper highlighting some increasing concerns about large passenger vessels. They were quickly getting bigger and carrying increasingly large numbers of passengers. Plus, many were advertising visits to some very remote parts of the world. Position papers like ours were used to garner support to create action at IMO. For this issue, IMO started a large passenger vessel safety subcommittee tasked with taking a risk-based, holistic look at safety onboard cruise ships and large ferries. The "holistic" part was important and would signal that the group shouldn't focus on small details of specific equipment requirements. For example, subcommittees already existed to establish requirements for lifeboats. This new subcommittee would look into why there was a set limit of no more than 150 persons allowed on a lifeboat. We found out quickly that no one knew why that limit was set. Would it make more sense if a bigger lifeboat was allowed? It would be more stable (and could possibly have a toilet).

As another example, we saw that cruises above the Arctic Circle were being advertised, but there were no additional safety requirements related to cold weather should there be a major problem onboard the vessel. When we reviewed trends and marketing materials, it also became obvious that cruise ships were traveling to some very remote areas in the South Pacific and around the world. Those voyages placed them well outside of any robust search and rescue capabilities that would be critical in a true emergency, not only for an individual passenger or crewmember that became ill or injured, but for the entire vessel should a major fire occur, or the vessel hit something and begin taking on water.

The large passenger vessel subcommittee brought together a real mix of countries and representatives. The Norwegians brought several members who all had strong opinions about what should be done. Other countries that routinely flagged (and taxed) cruise ships were there to make sure no undue burdens were placed on vessels flying their flag. We were there because many of the cruise ships sailed out of U.S. ports and carried large numbers of U.S. passengers. A delegation from Iran was also part of the group. They had several big ferries that carried large numbers of passengers and were, I believe, legitimately there to participate and learn what could be done to improve safety. Their challenge was that while the plenary sessions used interpreters who interpreted all discussions in real-time into six different languages, the subcommittees conducted their business in English. The Iranians struggled a bit with English and had trouble keeping up. The chairman did a nice job handling their questions and comments, which, because of the language issues, were always about five minutes behind the rest of the conversation. During the coffee breaks, I found the two Iranians from the subcommittee meeting at the tea table that I also visited and exchanged pleasantries. We kept the conversation related to the ongoing work, but by the end of the session, they said that they hoped that one day I could visit Iran as their guest. There was zero chance of that happening, but I reciprocated by asking them if there was any chance they could ever visit the U.S. Not surprisingly, their response was a lot of head shaking and frowns.

The session ended with the creation of a group that would continue the work by email between the sessions. I had the lead for the email work group and a short list of tasks to accomplish. Because we were just getting started on this holistic approach, we had a lot to do, but expectations were that we would work slowly and methodically. No one wanted a quick product that might be full of unintended consequences for the multi-million-dollar cruise and ferry industries. IMO never worked fast unless it was in direct response to an incident that shook things up. Little did we know that an incident was only a few months away.

I was at work in my office at Coast Guard Headquarters on Tuesday, September 11, 2001. I overheard someone say that an airplane had hit one of the World Trade Center towers in New York. I immediately thought it must have been a small private aircraft, but the atmosphere immediately got somber. We were glued to the television as the attack unfolded. While we watched the coverage with a focus on the World Trade Center in NY, we could see from our windows the smoke from the plane that crashed into the Pentagon. There were also reports of other explosions in Washington, DC. They all turned out to be false reports, but that "fog of war" created a lot of angst.

A few days later, I received an email from one of the Iranians who had been a part of the large passenger vessel safety effort at IMO and was part of the email working group. He expressed his hope that my family and I were okay, stated that the 9-11 attacks were terrible, and hoped that the perpetrators would be caught. It was interesting getting official email correspondence on my official Coast Guard email from a government employee in Iran. The first thing I did was report the email to my boss and the Coast Guard intelligence staff. I swapped one or two very generic emails with the Iranian, and that was the end of it. The exchange was just an odd thing.

A month after the 9-11 attack, I was asked to be part of an effort led by the Coast Guard Atlantic Area Commander, then Vice Admiral Thad Allen. The Coast Guard's port security responsibilities had just jumped from a low-level mission to the top priority, and the Coast Guard had serious resource constraints for port security. He asked our group to develop a tool that the field units could use to help prioritize their port security activities. We had a few sequestered days to develop something, and in his words, "He would ruthlessly implement a 70% solution."

We took a risk-based approach to the problem and were aided tremendously by some smart and experienced people with a contractor, the ABS Consulting Group, and some people from the Coast Guard's Research and Development Center. The tool we developed allowed the field units to review all the port risks and evaluate the relative level of risk for each. The riskiest would merit the most Coast Guard resources. The tool allowed individual ships in port or coming into port to be evaluated and compared to port facilities and other activities that might be adjacent to the port but that might not actually have a dock for ships to moor. It was based on the simple formula:

$$\text{Risk} = \text{Threat} \times \text{Vulnerability} \times \text{Consequence}$$

We tested the method and put it through its paces. We were pretty happy with the flexibility it offered. To ensure some level of consistency and repeatability, we generated some examples that we scored and rated. We quickly found that the tool's strength was as a relative risk evaluation. As long as the same people were doing the scoring and evaluating, there would be an acceptable level of accuracy and consistency.

Once developed, we needed to brief Vice Admiral Allen, which meant briefing some of his staff first so they could make sure we weren't going to waste the Vice Admiral's time if it was crap. The rest of the group wanted me to do the briefing. I prepped and practiced the briefing once, pitched it to the staff, then briefed Vice Admiral Allen. He was pretty happy with the product and said that he had asked for a 70% solution, and we had given him a 90% solution. I thought maybe he was just being generous, but I was quickly lined up to brief the Commandant and all of the flag officers (admirals) at Headquarters.

Then, a few weeks later, I briefed all of the Coast Guard admirals at a "Flag Conference." With everyone briefed and onboard, we got permission to launch it to the field units. To do that, we developed detailed instructions, specifying appropriate uses for the tool, specifying the limitations of the tool, and providing a whole lot of examples. The biggest limitation of the tool was that risk scores could not accurately be compared between units. If the risks were all identified and evaluated (or "scored") by the same group at a unit, the relative risk scores could legitimately be compared. However, people at different units might consistently skew the scores higher or lower than at other units due to their own inherent biases. As a tool that was intended to help each local field unit determine relative risks, it was great. Our concern was that senior staffs up the chain of command would want to compare scores from each unit to make manning and budgetary decisions. I'm not sure we convinced all of the admirals that the tool was not ideal for that use, but most seemed to understand.

It was all very cutting-edge for the Coast Guard, and we held several workshops and training sessions on the tool and other risk assessment methods. It was a great chance to make a difference and we made the most of it, laying the foundation for risk-based approaches later used throughout many Coast Guard daily operations.

Several years later, I heard that the Coast Guard boat forces had just developed and implemented a mandatory risk evaluation before every small boat launched, regardless of the urgency. The evaluation looked at the complexity of the case, the weather, the geography, the crew's experience, and the crew's level of rest/fatigue. If the resulting score was above a specified limit, they needed higher permission to get underway. It was a great approach, and I like to think that my team and I laid the framework for that.

The last year in that tour went by very quickly and as the spring of 2002 approached, I was ready for orders. I submitted my name for the commanding officer job at every Marine Safety Office coming open. There were only a limited number of them at the Commander level, and I suspected that my previous tour in Cleveland would land me back in the Great Lakes region. That would have been okay, but I really wanted a salt water port with a bit more action (and maybe more bearable winters). I was blessed when the selection board results were published and I got orders to Marine Safety Office Savannah, Georgia.

Despite being only 9 hours from the DC area, we only made it down for one weekend of house hunting and it turned into more of an area familiarization visit. Life was busy and time was valuable. I was moving fast, trying to wrap up several projects, make one more trip to IMO, help Peggy identify the best schools in Savannah for the kids, and take a trip with Peggy and the kids to Scotland and Ireland for the Irish Dancing World Championships. Our daughter, Haley, had qualified, and it was too sweet an opportunity to pass up. Before we left, Peggy had been looking online at houses in Savannah that were for sale trying to narrow down the choices. Our realtor was good and quickly understood what we were looking for and in which part of town. We ended up signing a contract for a house in Savannah while we were in Glasgow, Scotland. It was all done by numerous faxes sent back and forth at all hours of the day and night. The poor couple that ran our small boutique hotel was great and put up with the many late-night faxes and our comings and goings. It was a crazy time.

NINE

Commanding Officer and Captain of the Port (2002-2004)

We made the move to Savannah and it all worked out well enough, but we were reminded once again that things don't fall into place nicely unless you work hard and make sure they fall into place nicely. Buying a house in Savannah before closing on the house we were selling in Virginia, plus getting packed up and coordinating for the change of command in Savannah, was very hectic. It involved a few hundred details that we made fall into place.

I relieved a classmate of mine, Jim McDonald, at Marine Safety Office Savannah. The unit was in great shape, and everyone seemed exceptionally focused. That may have been in part because they were phenomenally busy. It was still less than a year after the 9-11 attacks, and there were a lot of Coast Guard reservists at the unit that had been called up to augment the active duty force. And their call-up had no termination date in sight. The reservists tended to be a few years older than their active duty counterparts, but were fully integrated into the unit and carried their weight.

The unit was conducting numerous security operations throughout the port each day, activities that were low or no priority before 9-11. Now, Coast Guard armed boarding teams from the unit were boarding ships every day offshore, conducting security sweeps on the vessels, and riding the vessels into port, providing security for them. The security boardings verified that the ship was not a security threat and then tried to ensure the ship would not be a target. That was all supported by screenings of each vessel done in the office related to risk factors for the ships and the relative risk tool that I had helped develop.

Even routine things had to be viewed from the perspective of port security. My first full day on the job, I attended a blessing of the fleet with the local Catholic Bishop. It was a pleasure to meet him, and I laughed when he told me I looked too young to be the Captain of the Port. I offered that he looked awfully young to be the Bishop! He laughed heartily as I suspected he would. Deep down, I was glad to see traditions like that being continued and glad it was my Bishop doing the honors.

That same day was the scheduled arrival of a new giant container crane for the port. The concern was the height of the crane. It was so tall that there was some uncertainty about it fitting underneath the bridge to get to the port. After a lot of calculating, our best estimate was that there would be two or three feet to spare between the top of the crane and the bottom of the bridge. It turned out okay with a bit more room to spare, but there were several tense moments. Secretly, I was thanking the Bishop for the blessing bestowed on the port, the vessels, the crane, and the bridge!

Marine Safety Offices were never, or maybe rarely, involved in operations that required Coast Guard men and women to be armed before 9-11. Now, Marine Safety Offices were sending out trained and qualified Coast Guard boarding teams and security teams to be the pointy end of the stick. It was exhilarating to me, but was not something I had done or was personally trained to do. What could go wrong?

Toward the end of my first week on the job, I called together all of the members of the boarding teams. It was a mix of active duty and reserve personnel, men and women, young and a bit older. Even though I had never done boardings like they were doing, I trusted in their training and "use of force" continuum. I wanted them to know that I understood the challenges and complexities of their tasks while boarding, searching, and providing security for huge commercial vessels with foreign crews bound for the port complex. I wanted them to know that I had their backs. So, I told them quite simply, "Don't shoot anyone…that doesn't deserve it." I explained that if there was a threat, I expected them to use their training and judgment and do their jobs. I further explained that if something happened, my first reaction would be to assume they were correct and did the right thing. Of course, there would be the required investigation, etc., but the starting assumption would NOT be that they had screwed up. The starting assumption would be that they did as trained and did their mission. I was never quite sure how they took my few words, but if they did need to pull their weapon, I wanted them focused on their training, not wondering if some jackass of a brand new Commanding Officer was going to look to throw them under the bus. I had seen enough senior officers who were "career first" and mission second. That just wasn't me. To me, if you focused on the mission, everything else would work out for the best.

I learned quickly that the Georgia Army National Guard were good people who were keen to protect their state and the U.S. The Adjutant of the Georgia National Guard, the general in charge, had assigned a company of National Guardsmen to protect the port in a crisis. They recognized that port security was a Coast Guard mission and that the Coast Guard had a lot of legal authority in that regard. They came to us and asked what they could do to help and where they could go in a crisis that would augment our actions with their capabilities. I loved that approach! We worked well with them and conducted a large-scale exercise with them that played well up their chain of command to the Governor. Aside from that, they were good people, doing good work, and it was an honor to work with them.

One funny thing with the Georgia National Guard was that someone in Washington, DC had decided that when the Soviet Union dissolved, the military services in the former Eastern Bloc countries would benefit from a close relationship with U.S. counterparts at several different levels. Thus, the Georgia National Guard was paired with the Republic of Georgia. That would normally have nothing to do with us. However, the National Guardsmen were desperate to find meaningful things to do with various military groups from the Republic of Georgia and regularly approached us to brief the delegations on Coast Guard missions. I was happy to help but managed to avoid joining them on the numerous delegation visits to the local Golden Corral, the Georgian military's favorite all-you-can-eat restaurant that served breakfast, lunch, and dinner. My whole staff and I felt a little bit bad for the poor major with the Georgia National Guard who drove the Georgian delegations around and had to join them for each meal at the Golden Corral, but none of us felt bad enough to join him or volunteer to take his place.

By late 2002, it was looking like the U.S. was going back into Iraq to fight. With the Army's 3rd Infantry Division located not far from Savannah, our port would be a major outload port for military equipment of all kinds. We started planning in earnest how to provide security for the port, security for the large numbers of military vehicles that would be staged to be loaded, and security for the ships that would be loading the vehicles and equipment.

My staff put together a great plan that addressed all of the threats. Part of the plan was a request for resources – mostly people and Coast Guard small boats, but also an additional volume of weapons and maintenance capacity. It was a solid plan, but my bosses at the Seventh Coast Guard District in Miami didn't like it. They felt the resource levels set a standard that was higher than what the Coast Guard could sustain at other expected outload ports if and when they had military outloads. But instead of just telling me the resources they would provide, along the lines of "You asked for four, but we can only give you three," they just wanted me to reduce my request. That just seemed strange to me, so I explained the tactics that we would use to protect and defend the ships and equipment, and how those tactics needed the requested resources. I felt that was especially the case since we were unsure how long the outload activity would last but expected it to be several weeks of round-the-clock operations. My bosses did not want to discuss details but kept suggesting that I ask for fewer resources. Eventually, I forced the issue. If they wanted a list of resource needs based on what I felt was best, they had it. If they wanted to order me to do the job with fewer resources, all they had to do was give me an order and I would salute smartly. The whole discussion just seemed bizarre to me. In the end, they ordered me to get by with less and sent the resources they wanted to send. We made it work because that's what Coasties do.

Now, Savannah, Georgia can be tremendously hot and muggy in the long summers. But winters can be a mix of cooler weather with occasional milder days. Except during the outloads in January-February 2003 when the temperatures dipped down to 10°F. I never would have guessed that a unit so far south would be ordering long underwear by the case to augment cold-weather gear for the boat crews. And we had a lot of Coast Guard boat crews and Coast Guard shoreside security forces brought in for the operation.

Right before we started the operation, we had an all-hands meeting. I wanted to address everyone so they could see who they were working for and so I could lay down a few expectations. I explained that what we were doing was important. Whether Saddam Hussein backed down or the U.S. went to war, the equipment being loaded was critically important. Just the fact that it was being loaded and sent overseas would potentially cause the Iraqi government to reconsider its options. Therefore, this security operation was big time and we all had roles to play to make it happen. I also noted that we had active duty and reservists from many other units but highlighted how we had established a clear chain of command structure for the operation. Every Coastie and every boat was assigned to a specific command element, had defined responsibilities, and was part of a defined chain of command. I expected them to respect and follow that command structure and raise any and all issues up that command and not back through their parent commands. With that, we got to work. It was a good, albeit cold, fight.

The outloads included some shrink-wrapped helicopters but mostly consisted of Humvees, trucks, and containers. In addition to providing security, we also had the traditional Coast Guard mission to ensure that the ship and vehicles being loaded were safe from the usual marine safety risks such as fires. We had Coast Guard men and women inspecting the equipment as it was being loaded and tracking progress on the ships to ensure there was no increased risk of fire or asphyxiation from exhaust fumes not being properly vented.

When the first ship was fully loaded, I got a concerned call one evening from my senior officers. The captain of the loaded ship had just completed the stability calculations based on the cargo that had been loaded, and there was a problem. The ship did not meet the stability criteria to safely get underway. This seemed odd to me because it was a capable, well-designed ship, and the cargo intuitively seemed to be well below the ship's weight-carrying capacity. I headed down there immediately. If this was going to be a problem, I wanted first-hand knowledge of the details. I met with the captain, he explained the situation, and we sat down in front of his computer with the stability program.

The Military Sealift Command had just sent its ships a new and improved software program to calculate the ship's stability. Stability programs perform the oftentimes complex calculations based on the ship's design and the exact cargo weight and locations on the ship. The locational information to be entered into the program included the longitudinal location for vessel trim purposes (whether the bow was lower or higher than horizontal), the transverse location for vessel heel purposes (whether the vessel was leaning sideways to the right or left), and the all-important vertical location that determines the ship's roll characteristics (and whether the ship will remain upright and not capsize). The captain showed me that the program indicated the ship was not stable and the vertical center of gravity was significantly too high. If accurate, some cargo would have to be offloaded.

I could tell that the captain was both nervous and frustrated. He was nervous because the indicated stability was bad and frustrated because it seemed counterintuitive to him as well. I had him show me how the cargo information was entered into the program. We went through it slowly. It took about 45 minutes of us calmly reviewing all the program data, and he figured it out. The new stability program that was sent to the ships required the vertical center of gravity of each piece of cargo (each truck, each Humvee, etc.) to be entered based on the height above the deck it was sitting on. The previous program required the distance to be entered from the keel (the bottom of the ship) to the center of gravity of the cargo. Not realizing the difference with the new stability program, the captain had entered the data just like before. Instead of the vertical center of gravity being entered as four feet above the third deck, it had been entered as twenty-four feet (its height above the keel).

With that realization, the correct data was entered, the program recalculated the ship's stability, and it was fine. A very relieved captain was able to get his ship underway and get the cargo heading to its offloading port in the Middle East. I felt bad for the captain. In the atmosphere of heightened anxiety and heightened security, and with millions of dollars of equipment being loaded, he had been sent new software to perform a critical function, and it did not come with any training. My contribution to resolving this "crisis" was not the application of advanced naval architectural knowledge but a sense of calmness and an analytic approach to problem-solving.

The Seventh District office in Miami was monitoring our operation closely and wanted a play-by-play on military ship arrivals, loading status, and ship departures. It was simple enough to keep them up to speed, although they raised holy hell when they saw that an "MSC" ship had departed without me notifying them. My boss called me up to chastise me after being yelled at by the admiral in Miami. I think they were all a bit embarrassed when I explained the "MSC" ship that had departed port was not a "Military Sealift Command" ship with military equipment but instead was a "Mediterranean Shipping Company" container ship with civilian cargo that regularly came to Savannah. It was the other "MSC." I didn't get much of an apology from the district staff, but I accepted the one I got and had a laugh at their expense.

The military outloads lasted for several chilly weeks and came to a successful conclusion. On a side note, the following November on Veterans Day, Savannah held their usual parade. I had the honor of sitting in the reviewing stand (back row with the less important people). I watched my son march with his drum as part of the Benedictine Military School band, as well as various other high school bands. Then we all heard a roar moving down the street towards us. It was the crowd cheering units of the 3rd Infantry Division back from Iraq. It was amazing to hear the yell from the crowd move along with the lead elements of the 3rd Infantry Division as they accepted some well-deserved recognition. Anyone who thought Savannah was an eclectic, artsy town that didn't go for overt displays of patriotism would have been sorely wrong that day. I proudly saluted the Army's 3rd Infantry Division that day.

In March or April of 2003, there was a worldwide outbreak of the SARS virus. It was officially named Severe Acute Respiratory Syndrome or "SARS" and produced a fever, then a cough, then frequently progressed to pneumonia. It was spread by respiratory droplets from close proximity to infected persons. It was more prevalent in Asia, but the Center for Disease Control (CDC) and everyone here in the U.S. were concerned. Anyone exposed or potentially exposed needed to quarantine for ten days. This was an issue because we got a lot of container ships that either came from Asia or had Asian crews, or both. And we boarded a lot of those ships using a very limited number of Coast Guard boarding team personnel.

I could not afford to lose a boarding team to the illness or a ten-day quarantine. It would mean that our level of port security would drop significantly. Guidance from Coast Guard Headquarters was entirely focused on what to do if someone was exposed, symptoms, treatment, etc. But there was no guidance on how to stay operationally capable. In short, their guidance sucked.

I did some research and stumbled across a U.S. law that required foreign ships to report sick crewmembers to the CDC prior to arrival at a U.S. port. None of us had ever heard of a ship reporting that, so I called the CDC in Atlanta. I asked them if they ever received reports from ships of sick crewmembers and was not very surprised to hear that they never did. That meant that if there were sick crewmembers onboard, we wouldn't know about it until the boarding team members were onboard the ship already and potentially exposed.

I also stumbled across another U.S. law that authorized the Coast Guard to assist with public health requirements. It probably dated back over 100 years to a time when immigrant ships would have to sit at anchor for a quarantine period. I had never heard of it, but it seemed to apply.

I debated approaching my boss in Miami with all of this. I had a problem, was trying to stay operationally viable, and had a possible solution. Seemed simple enough, but I knew the staff in Miami and suspected that they would find numerous ways to say "no." It was just their way, and they had the legal staff to support any "no" they wanted to provide. The best case was that they would understand and punt the issue up the chain. Maybe I would get some resolution in a few months. So, I took the Coast Guard option and issued a Captain of the Port notice requiring all inbound ships to Savannah to indicate to me whether everyone onboard was healthy or not and reminding them of their obligation under U.S. law to report sick crewmembers to the CDC. It seemed like a good solution that was not onerous for the ships and was within my authority. I had no intention of keeping my boss in the dark, but decided to ask for forgiveness instead of permission. I was exercising the Coast Guard option in the form of legal authority given to me as the Coast Guard's Captain of the Port.

When they found out, the staff at the Seventh District in Miami blew a gasket. The senior lawyer felt that I was exceeding my authority, but had a Lieutenant on the legal staff call me up to tell me that. That just pissed me off. I was a commander and did not take orders from a lieutenant. I asked him to solve my problem if he was going to tell me what I couldn't do. He had no answers and no solutions. That was typical from a very risk-averse staff.

Within a few hours, my boss and the senior lawyer were on the phone. Both were captains, and they were annoyed that I wouldn't accept their position as stated by the lieutenant. I repeated my operational concerns and tried hard to make them offer a solution to my problem. It was an ugly phone call and not my finest moment, but the senior lawyer eventually came up with a solution that would give me a heads-up on sick crewmembers. It was a different route to the same place. I rescinded my local requirement and implemented the suggestion from the District's lawyer. My crews appreciated as well that I was trying to keep them healthy and unquarantined so they could continue to perform the missions. It really was all about maintaining operational capability.

I felt totally vindicated many years later (and after I was retired) when the COVID pandemic was the issue in 2020, and Coast Guard Headquarters quickly developed and implemented policies that were almost exactly what I had put in place that had been met with angst by the Seventh District staff in 2003.

In late summer of 2003, we learned that the G8 Summit would take place in June 2004, about an hour from Savannah on the coast. The G8, or Group of Eight, consisted of the eight largest industrialized countries. In 2004, they were the U.S., Canada, France, Germany, Italy, Japan, Russia, and the United Kingdom. The meeting would involve the heads of state for each of those countries, plus numerous other invited heads of state. The meetings rotated between the countries, and it was the U.S.'s turn to host.

The more we learned about the summit, the more we recognized the complexities of the security challenges. The Coast Guard would be on tap to provide security on the water and coordinate with numerous other federal, state, and local agencies. The planning needed to start right away.

I asked Lieutenant Commander Alan Reagan to lead the planning effort. He was a reserve officer who had been called up right after 9-11 and had been at MSO Savannah since then. He was a Georgia native and an excellent officer gifted with a lot of common sense. At the unit, he was in charge of port security and boarding teams and was doing an excellent job. He had a huge role during the military outloads and deserved a lot of credit for its success. I was blessed to have him there.

He quickly developed a rapport with the two Secret Service agents who had been assigned to meet with us and provide needed input for our planning for the G8 Summit. Although not all of the details of the summit were locked in, the Secret Service agents shared key information as it became available and provided best-guess predictions for the details that were not yet settled.

Our first challenge was going to be geography. The dignitaries and the summit meetings were primarily on Sea Island, Georgia. It was accessible by one road from St. Simons Island, that itself was only accessible via one causeway from Brunswick and the mainland. That gave us two key bridges to secure, plus many miles of waterway around Sea Island and St. Simons Island and the offshore waters, too.

The second biggest challenge would be assessing the threats that we might have to deal with. There was a terrorist threat because so many heads of state would be tempting targets. Additionally, there was a threat from various anarchist groups, ANTIFA, black bloc groups, and protestors who were organized in varying degrees. Some of those groups had demonstrated the ability to show up in large numbers, be rowdy, be creative in their activities, and occasionally get violent.

The third biggest challenge would be the resource constraints from within the Coast Guard. We knew we would have to explain and justify in great detail every Coast Guard person, weapon, aircraft, cutter, and dollar we asked for. We would also have to be as spendthrift as possible and demonstrably so.

We recognized from the start that it was going to be a large-scale effort and that the logistics would be vitally important, especially in an area like Brunswick, Georgia, where we had very little Coast Guard infrastructure. There was a small Coast Guard Station in Brunswick in a small facility with just a few boats and crewmembers. That would never work as a staging area, command center, logistics base, or anything else beyond a boat ramp.

This multi-day event would require security all day, every day, starting before the dignitaries arrived and lasting until the last one had departed. We briefed my boss and the rear admiral at the District in Miami, Rear Admiral Harvey Johnson. They were very supportive, but I knew issues would arise when we started asking for resources. So, I briefed them early and often.

At one of the briefings, one of the captains on the District staff asked me why I should be the person to lead the Coast Guard effort. I suspected I knew what they were thinking. I was just a Marine Safety guy used to dealing with commercial ships and port issues, and they wanted someone with more operational experience in charge. I was a bit surprised because we were several months into the planning at that point, and I was also a bit annoyed. I replied that one of my several legal titles as Commanding Officer of Marine Safety Office Savannah was Federal Maritime Security Coordinator. That title came with legal authorities just like my other titles as Captain of the Port and Officer in Charge, Marine Inspection. I explained that this was a federal case that involved maritime security and was already requiring a heck of a lot of coordination. Therefore, as the Federal Maritime Security Coordinator, who else should be in charge? They bought that argument but assigned a captain from the District staff to be their point of contact and attend various meetings with me and my staff to better keep the rear admiral apprised. I figured that the liaison would do no work, be responsible for nothing, and just get in the way a bit. As it turned out, I figured correctly.

The way the Coast Guard shoreside operational units were organized in 2004, Coast Guard Groups owned the small boat stations, many of the patrol boats, and all of the search and rescue missions, plus some law enforcement missions, aids to navigation missions, and a few others. The boundaries of the Groups did not align with the boundaries of the Marine Safety Offices, which were based more on the volume and locations of commercial shipping activity. The Coast Guard Group boundaries were based on the workloads associated with the boating public and law enforcement needs. As a result, a big part of the planning and coordination involved Coast Guard Group Charleston led by Commander Jim Tunstall, and Coast Guard Group Mayport, FL led by Commander Mark Wilbert. Both of them and their staff were engaged and invaluable during the planning and the operation itself. They were good people, and we worked together just fine.

Additionally, we involved Coast Guard Air Station Savannah led by Commander Pete Troedsson. They flew H-65 helicopters and we knew they could be an important part of the operation. Several months into the planning of the operation, we were approached by the Secret Service agents assigned to us and asked if the Coast Guard would be willing and able to provide the slow-speed air interdiction component of the air security during the event. This would entail having a helicopter warmed up and manned in the Brunswick area ready to interdict any aircraft that entered the restricted airspace. And it would involve a second aircraft standing by in case the first had a mechanical issue when it had to launch. The interdiction was not an armed action, but more of a "see who it is and get them to land before we shoot them down." The Secret Service approached us when one of the State of Georgia agencies that had been asked couldn't decide if they wanted to take on the mission or not. They remained undecided for too long. We were the second option, and the Secret Service knew that we would at least give them a timely answer.

Pete Troedsson noted that the Coast Guard had never done that mission before, but he was willing to do it and there was still sufficient time to get the pilots and crews training properly. We would need to get permission from the District Commander in Miami before we could commit. The first call was to the District's Chief of Staff, who felt strongly that it was not something we should do. But when we briefed the Rear Admiral Johnson, the admiral gave an immediate and resounding, "Hell, yes!" With that, the planning expanded. The Coast Guard aviators were thrilled to have a bigger role in the operation, and it just felt like the right thing to do.

As we got closer to the start of the G8 Summit, news coverage ratcheted up and the protestor groups started to make some noise. One of the advantages we had was that the Summit was taking place in a small town. While that fact presented some minor logistics challenges for us, it meant the support in the local community to house, shelter, feed, and support the expected protestors was extremely limited. Plus, the Summit took place in early June, and South Georgia can be a brutally hot place to stand outside and protest or march.

While almost all of the Summit activity would be around Sea Island, there would be an event or two up the coast in downtown Savannah. We expected that some of the local Savannah population would likely be a bit more supportive of the protestors. Between the college students at the Savannah College of Art and Design and some of the eclectic residents and old hippies, there would potentially be a larger support network there than in the Brunswick/Sea Island area. We lived in Savannah and one evening, several weeks before the start of the operation I was out walking the dog when I was approached by a neighbor. I had seen him before but didn't know him. He was in his 50s with longish hair, was well dressed as he usually was, and as usual, had a glass of wine in hand while letting his dog get some exercise in a small park nearby.

He asked, and I confirmed that I was involved in some of the security for the event. He said he just wanted to let me know a few things. Savannah was his home, he loved it and was proud of it. He said he was also the kind of guy who would stop in the middle of the street and pick up a piece of trash to help keep Savannah looking good. He wanted to let me know that a certain woman who had been on the news offering her home and support to the G8 Summit protestors angered him. Further, he offered that if any of those protestors damaged his city of Savannah in any way, he would go over and burn her house down. I knew he wasn't serious, but I recognized his pride and willingness to protect his city. I chuckled and told him that if the city was damaged by protesters and he went to her house, he might have to get in line to set it on fire! He appreciated that and wished us good luck. Savannah does have a certain endearing charm.

About a month before the expected start date, we had an all-agencies tabletop exercise in Brunswick. The purpose was to establish coordination and communication. I drove down the evening before. I felt like I was coming down with something, but decided I would gut it out. That night in my hotel room, I was in a lot of pain in my lower right abdomen. I wasn't doubled over but couldn't stand up straight for long. It sort of felt serious, but sort of felt like it might pass. I tried to sleep, but I was too uncomfortable. It started to feel serious. Not being sure what to do, I prayed hard and asked Mother Teresa to intercede on my behalf and help me. A physical therapist that my wife worked with many years before had worked briefly with Mother Teresa in India and had mentioned her soft and healing hands, and how she never wore surgical gloves when treating patients. I asked Mother Teresa to take whatever was wrong with me, hold it in her healing hands, and surround it. I prayed a lot that night.

The next day, the pain was still there. I made it through the tabletop exercise and drove home, but by dinner, I was hurting. Peggy took me to the emergency room. You know it's possibly serious when you describe your symptoms and the triage nurse immediately takes you back with no waiting. After lots of poking, prodding, scanning, and head-scratching, the doctors felt it was probably my appendix. They scheduled me for surgery the next day and described it, in part, as "exploratory surgery." I thought they stopped doing that in the 1960s.

It turned out to be my appendix, but it had ruptured some months previously. The surgeon remarked that it was the first time he had ever seen a ruptured appendix where all of the ruptured infection stayed encapsulated in one place. Thank you, Mother Teresa.

Thinking back, I had had some acute pain in that area almost two years prior, but the pain went away, and the fever was gone in one day. It must have sat there for those many months festering. That explained why they also had to remove some adjacent intestine that had been seriously irritated from the infection pressing up against it all that time. It took a few weeks to fully recover, but I had work to do, and Mother Teresa was by my side.

The actual start date for providing the security was still up in the air as we got closer to the Summit's advertised start date. When it firmed up, the start was two days earlier than we had expected because President George W. Bush would be arriving early to take a few days off. We adjusted and were ready. Between Brunswick and Sea Island, Georgia, we had a 210' Medium Endurance Cutter offshore, two Coast Guard helicopters warmed up and mission ready, more Coast Guard boats than I had ever seen in one place, and approximately 1000 Coast Guard men and women. We stood up a temporary armory and coordinated boat maintenance teams to give us sustaining ability. Jim Tunstall, Mark Wilbert, and Pete Troedsson all had key leadership roles from start to finish and were amazing to work with.

When all of the Coast Guard small boats arrived, they were temporarily placed in a large hangar we had reserved. The crews lined them all up perfectly and one of the guys climbed onto a catwalk (and probably hung his butt over the edge) to take an amazing picture of the more than twenty boats, all in formation.

On a side note, I was promoted to captain just a couple of days before the start of the operation. The timing was coincidental but good. It was a big promotion, but any celebrating would have to wait. We had work to do.

In addition to all of the active duty and reservists, we had dozens of Coast Guard Auxiliarists involved. They are a volunteer group made up of civilians who offer their time and, for some, their personal boats as well, to assist the Coast Guard in any way possible. Traditionally, they conduct free, no-penalty boating safety checks, teach boating safety classes, serve as practice boats for hoists by Coast Guard helicopters, participate in search and rescue (SAR) cases, and help out in numerous other ways. Since this was a security operation and all of the boat crews would be armed, the was no role for Auxiliarists on the water. So, they drove vans and shuttled Coast Guard boat crews between the lodgings and all of the boat ramps and staging areas. They also delivered food and drinks to the crews. As some of the Auxiliarists were older, it was almost like parents and grandparents feeding their young Coast Guard kids. They were amazing and did it all for free.

I had established and formally broadcast the details of a security zone on the water that severely restricted any vessels, commercial or private, from getting close to the dignitaries. We used dozens of boats to enforce the security zone and, because of the size of the security zone, had to stage and launch those boats from several different places. I suspected that the security zone was the largest ever put in place by the Coast Guard, but there was no easy way to confirm that.

During the event, I was in the Multi-Agency Command Center (MACC) in a position to coordinate with all of the other agencies. Pete Troedsson graciously took the night shift while Jim Tunstall and Mark Wilbert ran the operational parts from locations closer to the boats. In the MACC, the Secret Service and the FBI had the lead, but there were other military components there, numerous agencies from the State of Georgia, and agencies from Glynn County and the local towns. It was a unique mix of agencies, authorities, and responsibilities, but we had been through enough planning meetings and tabletop exercises that we were comfortable working together.

President Bush stopped by the MACC on his first day in town to informally thank everyone for their efforts. It was a nice gesture and was appreciated by everyone in the MACC, all of whom had been engaged in the planning for almost a year. A few hours later, I was approached and invited to a reception and dinner the following evening with my wife. It was for the senior person from each agency involved in the security plus Republican Party donors, etc. I only had working uniforms with me, so I had to ask Peggy to bring a more appropriate uniform with her when she drove down from Savannah the following day.

That day was interesting. President Bush hired a local charter fishing boat that morning, which was initially swarmed by our Coast Guard boats as he drove into our security zone. Then we got the word that he was authorized by the Secret Service. Nevertheless, we closely escorted the President's boat. On the way back in, he stopped at a Coast Guard Aid to Navigation vessel that we had stationed as a remote logistics hub for our small boats. The President climbed onboard, thanked the crew, and gave them the fish he caught. A nice surprise for them.

That evening, Peggy and I attended the event. We didn't know anybody there, so we milled about smartly. The President and Laura Bush arrived, and a receiving line was immediately formed. When it was our turn to say hello and have a picture taken, the President saw that I was Coast Guard and said that he had met some of our guys earlier. I told him I knew about that and told him that the crew said, "Thanks for dinner." He did a quick double-take and started laughing. We posed for a picture, had a quick but nice dinner, then I went back to work while Peggy drove the 90 minutes home.

The Summit event was about as boring as we had prayed for. The few protesters to brave the heat were pretty tame. The one march that had the largest crowd walked across the long causeway in the full sun and nearly melted. By the time they got to the furthest point they would be allowed to reach, they welcomed the air-conditioned Georgia prison buses that took them back to the mainland and dropped them off. A few of the protestors wanted to be officially arrested at the demonstration and they were accommodated, but it was entirely calm (and sweaty, nearly heat-stroke-inducing, and nearly pavement melting). There were a few waterborne actions by the protesters, but nothing that posed a threat and nothing that needed any interdiction.

Just as the Summit was coming to an end, the funeral details for President Reagan were announced. He had passed away just before the start of the Summit, and his funeral was scheduled to take place shortly after the Summit. Several of the heads of state decided to stay to attend the services, which delayed their departure from Sea Island for another day or two. Thus, the security operation was extended.

As the operation was winding down during those extended days, one of the Secret Service agents quietly asked me to meet him at the side door in a few minutes. When I got there, he was rounding up a few of us to go meet President Bush, where he was staying. The President wanted to say thanks again. It was an informal photo opportunity, but I handed the President one of the patches that one of our Chief Warrant Officers (CWO) had designed for the operation. When I saw our "patch designer" CWO a day later, I told him that the President had one of his patches. He was really shocked and humbled, which I knew he would be. That's why I handed one of the patches to the President. The President didn't need it and probably had thousands of things like that, but I did it to surprise and honor our in-house patch designer.

Not long after the security operation ended, I had my change of command. The downside of getting promoted was that the Coast Guard wouldn't let me stay in Savannah as a captain in a commander's billet. I tried to push back on the orders to minimize the turnover at the unit, but it was wasted breath. I was ordered to a staff job at the Eighth Coast Guard District office in New Orleans. It was 2004, and we were going back to Louisiana, only this time to the Big Easy with our son halfway through high school and our daughter just starting high school.

The formal Change of Command as I left MSO Savannah was brief. Normally, the admiral from the District office in Miami would preside. But he was not available, nor was the admiral's Chief of Staff, nor was my direct boss available. I felt slighted when the District sent a reserve captain. None of us even knew what the guy actually did down there. I was even less impressed when he arrived in Savannah, missing a key part of his uniform, and made it my problem to solve. During the formal ceremony, I did my best to thank the crew and show my respect for their efforts. They were a great team and had done excellent work. Part of showing my respect for them was to keep my remarks very brief. I figured if I couldn't tell them what I needed to say in a few sentences, I wasn't worth listening to. I also thanked the local industry and port officials. They were key partners as we adapted to port security activities and overcame hurdles.

1. Christmas in Germany 1964 with my two sisters

2. A family outing circa 1967 with a scrapbook label showing my mother's sense of humor

3. Football at the Coast Guard Academy

4. Image of our Class of 1982 ring with our Class motto, "Audentes Fortuna Juvat" or "Fortune Favors the Bold"

5. CGA graduation. My mother pinned on one of my
Ensign shoulder boards

6. My father pinned on my other Ensign shoulder board

7. Peggy and I at my graduation

8. On USCGC STEADFAST in the drydock in 1982 while serving as an engineer onboard

9. Our wedding photo from March 1984

10. Graduation from M.I.T. I am on the far right, and my classmate, John Kaplan, is fourth from the left. We were all thoroughly soaked from the rain when this picture was taken.

11. The Atchafalaya River in Morgan City, Louisiana in 1991

12. Posing with our kids at Marine Safety Office Cleveland in 1997 at the Change of Command, where Ron Branch took over

13. Change of Command at Marine Safety Office Savannah in 2002

14. A picture with President Bush at the end of the G8 Summit in 2004

15. The Coast Guard Patch for the G8 Summit was designed by one of our CWOs

16. Peggy and I at an informal event before departing Savannah. My "I survived the G8 Summit" shirt seemed appropriate after months of planning and the extended operation.

17. With my parents at the Change of Command when I left MSO Savannah

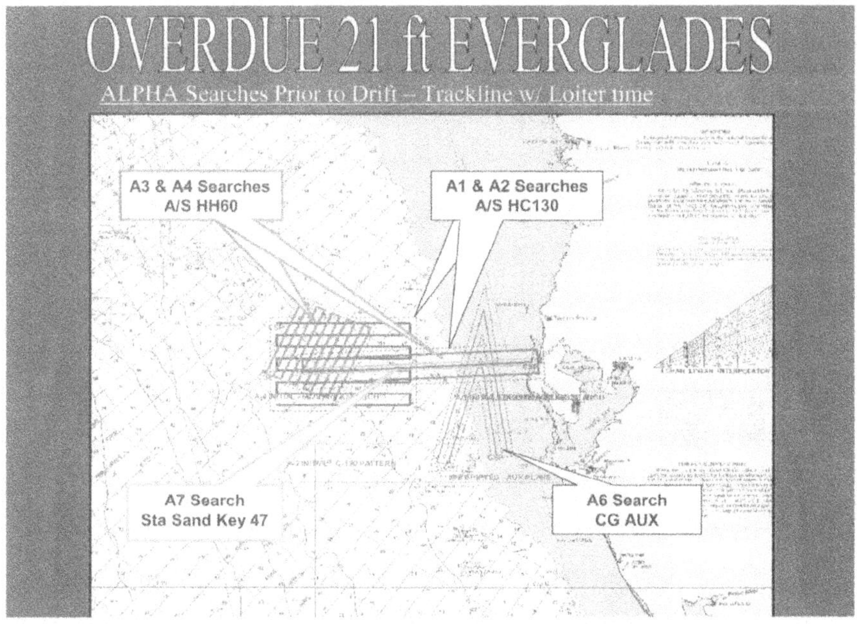

18. A page from the briefing book we provided to the families of the missing men during the large SAR case off Florida in 2009. This shows the initial, or "Alpha" search patterns.

19. Another page from the briefing book showing the composite of all of the search patterns.

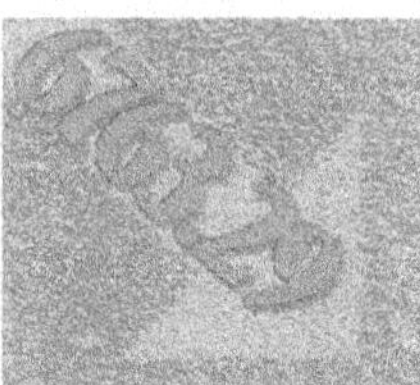

Down In The Dumps

The hits to the economy just keep on coming. On Monday:

>>> The Dow Jones industrial average fell below 7,000 for the first time since October 1997.

>>> The federal government agreed to provide an additional $30 billion to insurance giant American International Group, which also reported a $61.7 billion loss.

>>> HSBC Finance Corp. announced it will shut six Bay area lending offices as it pulls out of the U.S. consumer lending business.

DETAILS, Business, Back Of Sports

THE TAMPA TRIBUNE

and The Tampa Times

MARCH 3, 2009

TUESDAY

IMPROBABLE RESCUE

Photo by U.S. Coast Guard

One of the four boaters who went missing Saturday is found alive, but the search for his friends continues.

Top, a Coast Guard vessel rushes to aid Nick Schuyler who was transported by helicopter, above, to Tampa General Hospital. Schuyler was found dehydrated with dangerously low body temperature and is in serious but stable condition. The Coast Guard is adjusting the search for the other three boaters based on where the vessel was found.

By NEIL JOHNSON, njohnson@tampatrib.com
STEPHEN THOMPSON, sthompson@tampatrib.com
and KEITH MORELLI, kmorelli@tampatrib.com

TAMPA — The odds were low, almost nonexistent, and Marcia Schuyler knew it. Her son had been missing for two days in the open Gulf, and though the Coast Guard was continuing its search, his chance of survival dropped with the passing of each chilly, windy hour.

Finally, a phone call. The Coast Guard had found her son sitting on the hull of the upside-down boat about 40 miles southwest of Clearwater. He was dehydrated, scraped and bruised and his body temperature was dangerously low, but he was alive.

"I passed out," she said. "I went down."

Nick Schuyler, a former University of South Florida football player, was taken to Tampa General Hospital where he was admitted about 5:15 p.m. in serious but stable condition, more than 48 hours after he and

See MISSING BOATERS Page 6

- The four men left from Clearwater Pass early Saturday in calm weather, but heavy winds picked up through the day and the shoreside heavy, with waves of 7 feet and higher, Assistance alerted the Coast Guard early Sunday after the men did not return as expected.
- The Coast Guard had searched about 20,000 square miles of ocean for the boat by Monday.
- Monday afternoon, one of the missing boaters is found clinging to the overturned boat.
- The rescued boater is airlifted to Tampa General Hospital.

Tribune wire services; The Associated Press

COLD WATER IMMERSION

The Gulf temperature in the search area was about 68 degrees Monday.

Under 32.5 degrees: exhaustion in less than 15 minutes, survival time is 15 to 45 minutes.

32.5–40 degrees: exhaustion in 15 to 30 minutes, survival time is 30 to 90 minutes.

40–50 degrees: exhaustion in 30 to 60 minutes, survival time is one to three hours.

50–60 degrees: exhaustion in one to two hours, survival time is one to six hours.

60–70 degrees: exhaustion in two to seven hours, survival time is two to 40 hours.

70–80 degrees: exhaustion in two to 12 hours, survival time is three hours to indefinite.

Sources: University of Sea Kayaking, Maine Association of Sea Kayak Guides and Instructors

SEE STORY ON HOW TO SURVIVE PAGE 6

20. Local newspaper headlines of the rescue. Similar headlines made the news across the country.

everything. I can find contentment in a glass half full, "and that has made all the difference." Keep up the good work. I'll keep looking for your columns.
Michelle Keeney, *St. Petersburg*

Prepare for sea's danger

A little more than a week ago, a strong group of elite gentlemen set off confident in their knowledge and skill, but were overpowered by forces of nature. It is essential that the majority of us, who are no match for their strength, must take utmost care before setting out on the ocean in even the best of conditions.

One of the most important things a boater can do before leaving the dock is to provide a detailed float plan to a family member or friend. Float plans should indicate exactly where you're going and when you'll be home. This information is crucial to the Coast Guard should we need to find you. An accurate weather forecast for the area you'll be in is also crucial.

An Emergency Position Indicating Radio Beacon (EPIRB) is the single most important piece of emergency equipment a boater can have. EPIRBs are life savers. When an EPIRB gets activated, it sends a signal to the Coast Guard telling us you are in distress and providing us your exact location. A marine band VHF-FM radio can be extremely valuable, but has limited range. Additionally, cell phones become less effective the farther you are from shore.

Flares, strobe lights and other signaling devices allow us to quickly locate people who are already in the water and need help. We will search at night as well as during the day and anything that will make you more visible aids us significantly.

Finally, always wear a life jacket.

As we all continue to pray for Nick Schuyler's speedy recovery and for the families of Marquis Cooper, William Bleakley and Corey Smith, I hope the boating community will return to the ocean with both a sharper sense of nature's raw power and a renewed commitment toward personal safety. If so, I believe that some good can come out of this heartbreaking tragedy.
Capt. T.M. Close, commander, Coast Guard, Sector St. Petersburg

21. A brief editorial I wrote after the SAR case to try to generate some good from this case

22. At my classmate, John Kaplan's retirement

23. At the Change of Command, when I transferred from Sector St. Petersburg in 2010

24. With several classmates at the Change of Command in 2010. From left to right are John Weber, Pete Simons, Bruce Gaudette, me, Mark Vazquez, Kurt Hinrichs, and Bill McMeekin

25. Peggy and I with Commander Ray David of the Royal
Australian Navy at one of many official receptions in Hawaii

TEN

The Big Easy and the Western Rivers
(2004-2005)

The Eighth Coast Guard District is huge and encompasses the entire Gulf Coast from the Texas-Mexico border to the Florida peninsula near Tallahassee. It also extends up well into the heartland of America to include the Mississippi River and all of the rivers that flow into the Mississippi River. Collectively, they were known as the "Western Rivers" because that was how they were referenced when the country was young and just beginning to expand to the West.

I was pleasantly surprised to learn that I would be working with Ron Branch again, as he was in the process of transferring to the Eighth District staff as well. We had a good laugh about how the Coast Guard was clearly clueless because they were putting the two of us together again.

I had a lot to learn. I would be in charge of the Western Rivers and the six Marine Safety Offices spread along the Western Rivers. MSO Pittsburgh was at the start of the Ohio River in Pennsylvania. MSO Huntington was about 300 miles downriver from Pittsburgh in West Virginia, and MSO Louisville, Kentucky, was a further 300 miles downriver. MSO Paducah, Kentucky, was at mile marker 935, or 935 miles downriver from the start of the Ohio River in Pittsburgh. Paducah was also near where the Ohio River dumps into the Mississippi River. MSO St. Louis was responsible for the Upper Mississippi River all the way up through Minneapolis, Minnesota, and its tributaries. Last but not least was MSO Memphis, located on the Lower Mississippi River. While not falling within my area of responsibility, the Mississippi River system connects with the intercoastal waterway in Louisiana and runs along the Gulf Coast. It was truly a river system with emphasis on the word "system" – an interconnected transportation system with each river affecting and being affected by the others.

A little-known fact outside of the region is that the Coast Guard has such a presence up the Mississippi and Ohio river system and so far into the central part of the country. The explanation has to do with the huge volume of interstate commerce on the rivers due to the inland towing industry. That industry consists of blunt-nosed towing vessels that push barges up and down the rivers. The number of barges overall is staggering, and the volume of commerce is much larger than anyone would guess who doesn't work in or with that industry.

Some of the smaller towing vessels only push one or two barges at a time. On the Ohio River, the largest tows are limited to nine barges because of the size of the locks and the width of the river, which is narrower at its origin and only gradually widens as it approaches the Mississippi River. The largest tows on the Mississippi River can consist of thirty-five to forty barges routinely. The barges are mostly a standard size of 35' wide by 195' long. Some are open hopper barges that carry coal, gravel, grains, fertilizers, and other assorted bulk cargoes. Others are tank barges that carry crude oil, refined oil products, gasoline, diesel fuel, or any one of a thousand different chemicals. There are also specialized barges designed to carry liquefied natural gas (LNG), liquefied petroleum gas (LPG), and other cryogenic liquids. Some of the chemicals are downright nasty, like ammonia, chlorine, and some with huge chemical names that leave you scratching your head about both their pronunciation and their use.

The inland towing industry is a unique industry that contends with rising and falling river levels, ice issues in the winter, lock and dam maintenance delays, problems caused by groundings and collisions with river structures, plus the ever-changing rates they can charge to move specific cargos long distances. Many of the towing companies were and still are headquartered in or near Paducah, Kentucky, because of its central location, but others were located in various port cities along the rivers. They usually spoke with a common voice through the American Waterways Operators (AWO), an industry group that worked collectively for the inland towing industry. As such, they worked closely with the Coast Guard and the U.S. Army Corps of Engineers, which bears unique responsibility for the locks and dams on the rivers as well as various flow structures designed to channel the water into the middle of a river to reduce the buildup of silt. AWO did and still does a great job representing its members.

However, their goals did not always exactly align with those of the Coast Guard. We had a long-standing partnership to improve safety cooperatively and through non-regulatory means that I was very familiar with from my previous tour at Headquarters and the Prevention Through People initiative. However, AWO understand-ably wanted the Coast Guard to scrutinize their members' actions as minimally as possible, while the Coast Guard was responsible for navigational safety, pollution response, mariner safety, mariner licensing, incident investigations, and efforts to reduce the number of incidents. Although we shared several big goals, inherently, we were not always going to align with the AWO and the inland towing industry.

At that time, towing vessels were uninspected vessels. The Coast Guard had no legal authority to inspect them like the authority we had for large ocean-going ships, passenger vessels, tank vessels, and many others. But we could shut down a vessel if it were inherently unsafe. That was admittedly a case-by-case judgment call that would be subject to scrutiny and subject to appeal. And since the Coast

Guard issues licenses to the towing vessel captains and pilots, an investigation into a marine casualty, such as a grounding or collision with a bridge or lock structure, could result in a suspension of a mariner's license.

The Commanding Officers at the six MSOs up the river system reported to me. I would also be the Eighth District's primary liaison with the inland towing industry and the American Waterways Operators. I had a lot to learn, but was sure of a few things. First, I would try harder to support those units better than I felt I was supported by the Seventh District when I was in Savannah. Secondly, I would not let the District Commander, i.e., the admiral, and the District staff get fixated on the Gulf Coast units at the expense of the inland units, even if the inland units were smaller and not as exciting.

In mid-September 2004, Hurricane Ivan hit the Gulf Coast. The Eighth District stood up our incident management team, dividing much of the staff into two twelve-hour shifts. We served as command and control for the field units while supporting the District Commander, Rear Admiral Bob Duncan. Ivan was initially projected to make landfall near New Orleans, but we knew there was some unpredictability in terms of course, direction, and wind speed.

With New Orleans within the predicted range of potential landfall locations, an evacuation was recommended. As was standard practice, the Coast Guard issued evacuation orders for our families and identified a Navy facility in central Mississippi as a destination. Peggy, the kids, and the dog loaded up and hit the road, only to grind to a halt in a massive traffic jam as everyone else in New Orleans tried to get out of town. It took them hours to travel just a few miles. Then we got the word that the Navy facility was full and was turning away additional families. That was not good.

As the junior captain, I had the 7:00 pm to 7:00 am shift at the District office in New Orleans. Then, per our plan, my team and I hopped on a Coast Guard aircraft before the hurricane made landfall and flew to St. Louis, MO, where we had space set aside in a federal building. We would exercise command and control from there in case New Orleans got hammered by the storm. Shortly after landing in St. Louis late that evening, I got a call from Peggy, who was driving but didn't know where to go. It didn't help that I was in a hotel room in St. Louis while she and the kids were trying to sleep in the car in the middle of nowhere with the dog in full protection mode and barking at everyone and every sound. I asked a favor from a guy I had worked with at Headquarters, who was the XO at MSO Memphis, TN. Wayne Arguin, currently a rear admiral, graciously offered to put up my family. I vectored Peggy to Memphis, and it worked out, but it wasn't easy for her and the kids.

The storm missed New Orleans altogether and instead made landfall in Alabama as a Category 3 hurricane. It caused a lot of damage, but mercifully less than if it had remained the Category 4 storm it had been just a day or so earlier.

Back at the District, we reviewed what had worked well and what had not. I was pretty vocal about the need to do better by our families and to have better arrangements in place in advance. Changes were made over the next several months. Little did we know how much those would be needed the following year.

ELEVEN

Flood Stage (2005)

In what I recall was the late winter or early spring of 2005, the weather up and down the river system was bad with lots of rain and the start of the spring melt. The river levels were very high all along the Ohio River, as well as the Upper and Lower Mississippi River, and would not be dropping anytime soon. It was not the worst ever by a long shot, but the conditions on the river were bad.

In simple terms, it worked like this: several days of heavy rains across the area from Pittsburgh to Cincinnati would create localized flooding issues, and all the rain would quickly drain into the Ohio River. The river level in that area would rise and continue to rise until all that rainwater flowing downhill slowed. That large slug of water would move down the Ohio River downstream from Cincinnati and raise the river levels downstream even though it may not have rained there. That slug of water would be largely predictable in height and duration. As that water continued downstream and flowed into the Mississippi River, it would raise the river levels on the Mississippi River. The river rise would be lower in amplitude because the river was wider, but longer in duration, as the slug would lengthen as it moved downstream. Eventually, New Orleans would see a rise in the level of the Mississippi River as that slug of water passed New Orleans and continued downstream into the Gulf of Mexico. The situation gets substantially more complex and harder to predict if, as in this case, there were also heavy rains falling along portions of the Mississippi River basin, and the snow was melting in Minnesota, Wisconsin, and Iowa, and adding more water to the tributary rivers in addition to that from the rain.

High river levels always created the most complex and dynamic conditions through which to navigate because the current was faster and seriously less forgiving. It was not long before a few towing vessels lost control of their tows and hit various bridges and critical parts of the locks and dams. Several of the incidents stopped all traffic on parts of the river until the local issue was resolved. The local Coast Guard Marine Safety Offices were responding to each incident and would investigate each incident, but the incidents were happening frequently. They were also causing a lot of damage and a lot of delays to other tows trying to move up and down the rivers. Some of the damage to the U.S. Army Corps of Engineers' locks and dams put the locks out of commission for many days. Some of the delays caused, for example, chlorine supplies used by cities for municipal water systems to run dangerously low.

Finally, we all had had enough. My boss, Rear Admiral Duncan, told me to figure out how to stop the incidents from happening and do it quickly. The towing industry had long advocated that they knew what they were doing, had been doing this for decades, and didn't need the heavy hand of the Coast Guard getting in their way. They had invested a lot of time and effort with Coast Guard personnel who were new to the rivers and taught us about the complexities of the river system and navigating large tows. We were willing and active learners, but did not forget our basic roles and functions related to maritime safety.

I organized a quick conference call late that same day with all six of the MSO Commanding Officers, the U.S. Coast Guard Captains of the Port. My tasking to them was clear: do whatever it takes to prevent any more of these incidents. I told them to put in place any and all restrictions for their respective portions of the rivers and do it immediately. We were taking the Coast Guard option since the alternative of letting the towing industry continue on its own was just not working.

As expected, the towing industry was not happy about such widespread Coast Guard intervention, but they had a difficult time arguing that they had this under control. They did not. We had to get through this specific period of high river levels, but I wanted to set the stage for the next time the rivers rose and the next time the rivers dropped significantly. And we all knew that with the cyclical nature of the rivers, there absolutely would be another period of very high river levels.

In our discussions, I came up with the idea that might work. What if we established actions based on predetermined river levels for each geographic stretch of each river? Because the rivers varied so much along their length, and because the hazards varied accordingly, the predetermined actions and levels would have to be different for different sections of the river. Getting an agreement on what the trigger points should be and what actions would be best would take a lot of work and would involve the various Coast Guard MSOs, the U.S. Army Corps of Engineers, and numerous towing companies along each river. But it inherently felt like a good fight and a worthwhile effort.

Convincing Rear Admiral Duncan was fairly easy. He recognized that we would be trying something that hadn't been done before and that there would be a lot of angst throughout the towing industry. But he also felt that the status quo was not sustainable.

I kicked around the concept with each of the MSO Commanding Officers. The burden of the details would largely fall on them. We had some good discussions and explored several options about how to make it happen. We all agreed that I had to get AWO onboard first, and that meant getting the owners of several of the larger towing companies onboard.

Some of those initial discussions with AWO were spirited. But I think they appreciated that we were directly responding to a series of incidents that they had created. I asked them to be part of the discussions about what the triggering river levels should be and what the appropriate actions should be when those river levels were reached or about to be reached. I wanted them to be part of the solution because I knew that the temporary restrictions put in place by our Captains of the Port would possibly be too stringent in some cases and insufficient in others. I needed the towing industry to weigh in, and I valued their input. And I also needed the U.S. Army Corps of Engineers to be part of the solution because they understood the river levels and the hydrology of the rivers. They also owned and operated the locks and dams, so they had a lot at stake.

It took a lot of convincing, and I had several towing company executives tell me that I was crazy and that it would never work. I took that as a challenge.

I think at some point, the senior AWO staff saw that this idea was not going to go away and that it would be better to get onboard and be part of the solution. That's when the discussions started in earnest. And that's when we started to make some progress.

We agreed to divide the river up into defined segments with predetermined triggers and actions associated with the rising (or falling) river level and the latest river level predictions. We also agreed that the first trigger and action would always be simply a conference call involving the local Coast Guard, the local Army Corps of Engineers, and as many of the local towing companies as wanted to participate. The call would review the current river conditions, review the forecasted river conditions and the expected timetable, and discuss the next trigger point and actions. We also agreed that the trigger points and mandated actions would be reviewed periodically and could be adjusted. It needed to be a living document that evolved as the river system evolved and the towing industry evolved. It was clear this would not be a quick effort nor a simple one.

We set up working groups in each river segment to identify the trigger points and actions. For example, they might agree that in a specific part of the river, when the river level rose to a certain point, only tows under a specified number of barges would be allowed to transit, or only tows with a horsepower-to-barge ratio above a specified level would be allowed to transit, or that an assist towing vessel had to be used for the transit. The trigger levels and forecasts were key because the river would be very different if it was peaking at ten feet above flood stage compared to reaching ten feet above flood stage but forecasted to continue to rise. Managing the flow of traffic required recognition of the river's current condition as well as an understanding of what would be expected in the following few days.

Once the local Coast Guard Captains of the Port started to sit down in the working groups with the towing industry, the towing company representatives started to recognize that we weren't trying to be draconian and weren't trying to eat into what profit they were making. We kept emphasizing that the goal was to keep the waterways open for everyone all the time by avoiding those incidents that blocked a river or damaged the infrastructure and created a blockage or major choke point. We reiterated that point early and often.

Just as common as periods of high water were periods of low water. They happened every year, but the length and severity were entirely dependent on rain and snow that did or did not fall weeks or months before, and many hundreds of miles away. What is not common knowledge is that as the river level drops, the current slows down. As the current slows, the water cannot hold the same volume of silt that was entrained previously. The silt drops out of the water and builds up in various locations in the river, some of which are predictable and some of which are not.

The result is that low spots will form in the river with insufficient water for the barges and towing vessels to smoothly transit over. Thus, they run aground and get stuck. Coast Guard buoy tenders work their butts off adjusting buoy locations to mark "good water," but when the river is falling, it is nearly impossible to adjust buoys everywhere at the same time. A towing vessel and barges that get stuck in conditions where the river is still dropping could be stuck for a while. It could also block the river completely for barges with shallower drafts until it is freed. While those groundings tend not to create a lot of physical damage to the barges, towing vessels, or infrastructure, the real damage is in the delays to cargo. Therefore, we needed to create trigger points and actions for periods of low water with the same idea of keeping traffic moving.

I'd be remiss if I didn't point out a few noteworthy things the inland towing industry did as a whole. As happened on occasion, a river blockage would cause a queue to form of tows both upriver and downriver, waiting for the issue to be resolved. The industry would get together and prioritize the tows to go first when the river was again open for transit. The queue was rarely "first come, first go" and was usually based on the demand for the cargo at its destination and the nature of the cargo. For example, barges loaded with coal that were being stockpiled at a power plant might be a lower priority than a barge carrying a chemical needed for a manufacturing facility that had limited storage capacity at the facility. Industry representatives worked this out among themselves, and we were happy to let them. Unrelated, while the towing companies may have been suspicious or skeptical of almost everything the Coast Guard did or wanted to do, they always remained willing to talk to us and listen, and that rapport we built was much appreciated.

It took almost a year to get the entire series of trigger points and actions worked out and documented. We called it the "Waterway Action Plan" and had a signing event with the Coast Guard admiral from the Eighth Coast Guard District, the general from the U.S. Army Corps of Engineers, and the chairman of AWO. It was a good day and the culmination of a lot of hard work. We took the Coast Guard option, used our authority to force everyone to the table, and made it work.

One of the same towing company executives who previously told me the idea was crazy was the first to say that he had spoken too soon and that it might just work. That was a measure of success in itself. The real proof of success is that the Waterway Action Plan is still in place and is being used many years later. It truly was a good fight and worth the exercise of Coast Guard authority.

TWELVE

Katrina (2005)

In late August of 2005, the Eighth Coast Guard District again stood up its incident management structure for another hurricane that turned out to be anything but another hurricane. Hurricane Katrina formed quickly and grew massively. Once again, I had the 7:00 pm to 7:00 am watch for the District, and once again, families were issued orders to evacuate. After the previous summer's hurricane, everyone was better prepared. The state implemented contra-flow on the interstates, allowing the inbound lanes to be used for outbound traffic and people evacuating.

Peggy, the kids, and the dog hastily packed up and headed to Naval Air Station Meridian in Meridian, Mississippi. As soon as I got off watch, I buttoned up the house as best I could and quickly hung and secured the hurricane shutters. I crammed as much outside stuff as I could into our garage, then gathered up some uniforms and personal gear. Then I hurried back and took the next watch as scheduled and started the next few days' marathon, dog tired.

When my duty section reported for watch, we brought with us our bags so we could deploy from there directly to St. Louis when our watch ended the following morning. There was still some uncertainty about where the storm would make landfall, but it looked more and more likely that New Orleans would take a hit. Rear Admiral Duncan planned on heading to St. Louis with us, but I talked him out of it. Once we were in St. Louis, it could be difficult for him to quickly access a Coast Guard aircraft. Instead, I suggested he relocate to Coast Guard Air Station Houston, where he would have a variety of Coast Guard aircraft and crews available for his immediate use. That turned out to be a very good recommendation.

When we got off watch, a member of the oncoming duty section drove us to meet a Coast Guard C-130 that flew us to Scott Air Force Base just outside of St. Louis. By 7:00 pm that evening, we were back on watch in St. Louis. The watch section we left behind in New Orleans, led by my friend and classmate, Captain Pete Simons, finished their watch there, then drove out of town to meet up with the Coast Guard Sector New Orleans command in Alexandria, Louisiana. I may have been on the last aircraft out of New Orleans, but Pete and his watch section may have been the last vehicles out of New Orleans. I think by the time they left, everyone who could evacuate had done so already.

On the District's Incident Command watch, we were in contact with Coast Guard Sector New Orleans and Coast Guard Sector Mobile, as well as the Coast Guard Air Stations. They would run the operations in their areas, and we would provide and coordinate resources as needed. We were also in contact with the next level up the chain of command, the Atlantic Area staff, who stood up their own robust incident management team. Resource requests that we couldn't provide would be punted to the Atlantic Area.

As was Coast Guard standard operations, we only kept assets and people in the path of a hurricane as long as it was safe to do so, then our people and equipment, boats, aircraft, cutters, etc., would evacuate to safety, then get ready to swoop back in and get to work as soon as the weather allowed.

Katrina became a Category 5 hurricane but made landfall in Louisiana in the early hours of 29 August 2005 as a strong Category 3 hurricane, and it raged all day. It would cause over 1300 fatalities and flood 80% of New Orleans and many of the surrounding parishes. But we didn't know that as it was unfolding. For most of the day on the 29th, all we could do was watch.

Late in the day, Rear Admiral Duncan left Houston on a CG HU-25 jet, a Coast Guard variation of a small business jet. At the very edge of flying conditions, the plane flew over New Orleans and the Mississippi coast while there was just enough sunlight left to see what was happening. As soon as the plane landed, he convened a call with his senior staff, including me. I was back on watch by then, only in St. Louis.

He said that he saw the storm surge pouring back out of Mississippi along the coast with large amounts of damage. He also reported that New Orleans was flooding and that some people were on their roofs trying to stay above the rising water. They needed rescuing. He said he wanted "to darken the skies with Coast Guard aircraft." Yes, sir.

My next call was to the Atlantic Area. I relayed what Rear Admiral Duncan had reported and his orders. They told me that if I had a request for resources, I needed to be specific. That sort of pissed me off. So, I said, okay, we want it all. We want every Coast Guard aircraft, pilots to fly them, flight mechanics, lots of rescue swimmers, and enough deployable helicopter support kits to maintain them for the foreseeable future. And we wanted them now. I told them to get everything moving to Coast Guard Air Station Mobile, which we knew by that point was intact and operational.

One of the great things about Coast Guard aviators is that they do not need a lot of direction. They are used to operating independently and are experienced at changing the plan mid-flight as circumstances change. The Coast Guard aviators and aircraft from Eighth Coast Guard District units responded right away, and the rest of the Coast Guard aviation community followed. Within a few short days, every Coast Guard Air Station in the U.S. had sent pilots, flight crews, swimmers, and support crews. And many of the Air Stations outside of Alaska and Hawaii sent helicopters.

We learned pretty early that Air Station New Orleans was intact but without power. No power meant no refueling. I told my logistics guy in my watch section to contact MacDill Air Force Base and request remote fueling capabilities in the form of portable bladder bags, pumping equipment, and a few technicians to operate it. It was important to get Air Station New Orleans functional to shorten the distance the aircraft had to fly to refuel and swap crews and pilots.

The Coast Guard aviators did great work in New Orleans. They didn't train for urban rescues, but in the immediate aftermath of the hurricane, people needed help, so the aviators got busy. They created and adapted techniques and equipment as needed. They used cliff rescue techniques to get their rescue swimmers into high-rise apartment buildings. Other rescue swimmers began using the helo's crash axe to chop through the roofs of houses to rescue people trapped in their attics. Then, they graduated to fire axes with longer handles taken from the buildings at the Air Station. Then they shifted to gas-powered rotary saws. They did hoist rescues while fighting the rising smoke from building fires. The rescue swimmers swam through some nasty water to reach trapped people. They did it all without the usual landmarks and flight plans. They deserved all of the recognition they got for it. But they were not alone.

Captain Sue Englebert was the Commanding Officer at MSO St. Louis and soon to be the first Commander, Coast Guard Sector Upper Mississippi River when the Marine Safety Offices were later combined with the Coast Guard Groups. When she saw that New Orleans was flooding, she gathered her department heads and senior Coast Guard Reserve personnel. She tapped into their hunting and camping knowledge and figured out how to send a contingent to New Orleans that could help and be self-sufficient. Significantly, her unit had flood punts, small shallow draft aluminum boats with small outboard engines that were trailerable.

They had the flood punts to respond to flooding throughout the upper Midwest when rivers overflowed their banks, which was an occurrence that happened frequently enough. By the time I asked her how she could help, her teams were already heading south. Sue was just exercising her own Coast Guard option. That was good on her!

I was all in as we tried to support and coordinate with the units in New Orleans. The search and rescue cases were fast and furious by air and by boat. My twelve-hour watches were fast-paced and frenetic as we responded to requests for resources and requests for updates while handling one unique problem after another.

A very minor issue but a telling one was that several people assigned to the District staff had been away from New Orleans on leave when we stood up our incident management teams and later evacuated. Several of them joined us in St. Louis directly from being on vacation and without their uniforms. Others were in the middle of transferring to the District staff but had to join us in St. Louis before they ever made it to New Orleans. Several of them were without uniforms or with minimal uniforms. The units responding to the storm in New Orleans and its aftermath were trashing uniforms quickly and needed more. Coast Guard enlisted members are issued uniforms initially and then, thereafter, are given a small allowance to purchase replacement uniforms as needed. Officers always paid for their own uniforms. To their credit, the Coast Guard uniform distribution system recognized that the situation was not business as usual and sent large batches of uniforms to the units in New Orleans and smaller batches to us in St. Louis. No one was ever charged, and no accounting was demanded. It made me proud that in a time of need, the red tape was pushed aside.

In another example, we learned that the Coast Guard's civilian pay system, at least for the entire Eighth District, ran through computer servers that were in New Orleans. They were without power and likely flooded, as they were in a ground-level computer room somewhere. On the military side, we knew we would still be paid regularly, but the civilians worked off a different system that required timesheets to be submitted every two weeks. If they did not submit a time sheet, they were not paid. Recognizing that submitting time sheets would be difficult for our civilian employees, the Coast Guard announced that they would continue to pay all civilians based on their last paycheck before Katrina hit. While it made perfect sense, it was still amazing that they made that call and did so quickly.

In one last example, our families that evacuated from the New Orleans area were officially given evacuation orders. That entitled them to a food allowance and lodging reimbursement while they were in that evacuation status. Early in the response, an official announcement was made that the food allowance and lodging reimbursement would be extended well beyond the usual time period. They recognized that some families had no homes to return to in New Orleans and the surrounding area, and that those whose homes were intact would still not be able to return any time soon because most were without power, and there was little or no supporting infrastructure in the area. It was, of course, the right call, but the decision was made quickly and applied broadly. That relieved a lot of anxiety and allowed a lot of Coast Guard men and women to better focus on the work that needed to get done, knowing the financial hardship on their families was lessened.

The operational response was amazing. Some actions were downright heroic. Others were just good Coast Guard people doing their jobs well. Those three quick examples of letting the need drive the bureaucracy also made me proud. That was my Coast Guard doing the right thing.

My first watch in St. Louis began late in the evening. I addressed the team with a question – Does everyone know where their families were? Most of us did, and those who didn't were tasked to find out as a priority before they got to work. Then I asked my logistics guy to find a local all-night delivery place so we could get some food. Then we went to work.

Many of the watches were a blur of activity. A year later, when I moved a stack of papers that turned out to be my notes from those watches, I stopped and took a quick look. To my surprise, many of the actions and phone calls that I thought happened over several days actually all took place on the same day during one watch. Your memory just plays some tricks on you regarding the timing.

Generally speaking, information never came in as fast as I wanted, and never got compiled as quickly as I needed. I desperately wanted to be onscene in the middle of the action, even though I knew some of those guys were getting burned out. I had seen a few others crumble under the uncertainty, stress, and fast pace. They were only built for steady-state operations. I wanted the fast pace and action, but had to settle for my role. It was more than a little frustrating.

A few weeks into the extended operations, we downsized our duty sections. The crisis phase was over, and with it, the need for a robust 24x7 incident management team ended. I went back to working days, albeit from a temporary desk in St. Louis.

I was able to make a quick visit back to New Orleans to check on our house and grab a car and a bunch of stuff that Peggy and the kids needed. They had only stayed one night in Meridian, Mississippi, before Peggy recognized that the hurricane would pass directly over them. They drove to Sarasota, Florida, to stay with Peggy's family initially. Once the emergency phase had passed and it became obvious they could not return to New Orleans, they relocated to Savannah, Georgia, and she re-enrolled the kids into the same schools they had attended before we moved to New Orleans. Friends in Savannah had a rental property house they could stay in, and friends helped them with furniture and kitchenware, etc., that made it work.

I drove from New Orleans to Savannah, had a quick visit with my family, then jumped on a plane back to St. Louis. Another hurricane was heading to the Gulf Coast. Hurricane Rita made landfall in western Louisiana not far from the Texas border. It was a big storm and even caused New Orleans to re-flood in large areas, but mercifully didn't cause as much damage or result in the fatalities that Katrina caused.

With the search and rescue over from hurricanes Katrina and Rita, the focus shifted to the numerous oil spills in the area. Many were from oil well production facilities scattered throughout the bayous and the areas just offshore. Some of the spills were from oil storage tanks that had been blown or washed off of their bases, with the complete loss of all the contents.

After a few weeks of exposure to the Louisiana sun and heat, all of the components of the oil that could evaporate had done so already. What was left was the thicker, gooier oil that clung to marsh grasses. The Coast Guard oil spill responders working for Captain Frank Paskewich, the Coast Guard Sector Commander, asked for permission to ignite the oil and burn the marshes. That was a proven technique when done carefully, but burning it required permission from numerous other federal and state agencies. I participated in many of those conference calls to get everyone's blessing. Those were some of the shortest conference calls I have ever participated in. A few of them lasted a mere five minutes, and most of that was the roll call. In that crisis climate, all bureaucratic nonsense and inter-agency issues were set aside, and every agency was helpful and focused.

In addition to oil in the marshes, some oil was still on the water in more open areas. Paskewich's responders coordinated with spill response companies and the various federal, state, and, in some cases, Parish (the "county" in Louisiana) representatives. Oil spill response vessels came in from all over the Gulf Coast to assist. Some of the oil was collected and recovered. Some of it was collected using special "fire boom" and burned in place.

The issue that became critical was that the oil spill response vessels working off Louisiana were no longer available in the other ports they came from, especially the Houston-Galveston area. Oil tankers and product tankers carrying refined petroleum products were required to have oil spill response companies and equipment available within limited time periods should there be a spill in that area. Because so much of the equipment had responded to the spills off the coast of Louisiana, the tankers in the Houston-Galveston area were not in compliance with their oil spill response plans and, therefore, could not load or offload their cargo. With the ports in the greater New Orleans area still not operational, it was important for the region and the nation to get oil flowing again in the greater Houston area.

We devised a solution that hadn't been tried before. What if we gave the Coast Guard Captain of the Port emergency authority to accept temporary alternatives to the type and amount of spill response equipment? In reality, the big spills off the coast of Louisiana needed the larger ocean-capable oil spill response vessels. The smaller, near-coastal vessels and equipment had stayed in Texas. With the good weather in Texas, there was no real need for ocean-capable spill response vessels in that area. We needed to give our Captains of the Port the authority to accept a quantity of near-shore vessels and equipment as suitable alternatives to the ocean-capable platforms and equipment until they were done cleaning up oil off Louisiana.

It made sense. We just had to convince the Secretary of Homeland Security to issue an emergency regulation allowing that! He was just a hundred layers up the chain of command from me, and there would only be a few dozen legal offices that would have to sign off on the emergency regulation. It took two full days of emails and phone calls from my small, windowless, temporary office in St. Louis to get it done. I wasn't acting alone. The people in the Coast Guard chain of command all recognized that this was a good solution to a real problem that needed to be solved immediately, so oil could begin flowing into and from the U.S. Gulf Coast again. We just had to get their attention long enough to explain the problem and our proposed solution. I always suspected that a small part of the reason they were so quick to buy off on the solution was that it placed all the risk on the local Captain of the Port and none on anyone up his chain of command. Coast Guard Headquarters not only signed off but did critical work getting the issue before the Secretary at DHS and getting the wording just right for an emergency regulation. It was another victory for common sense and another example of the bureaucracy not getting in the way of doing the right thing. We used the Coast Guard option again.

The Coast Guard remained extraordinarily busy throughout the region well after the hurricane made landfall. Lieutenant Commander Scott Calhoun (now a retired captain), who had been on my staff at Headquarters, and Joe Myers developed a risk-based tool to help the Coasties in New Orleans prioritize vessel wreck removals by systematically comparing the relative risks of each. Oil spill clean-up work was still ongoing and would continue for many months. Opening the waterways was a priority but required salvage work, side-scan sonar equipment to determine that the waterway was unobstructed below the surface, and then replacing missing or destroyed aids to navigation. The workload was huge, and the Coast Guard surged large numbers of additional Coasties to assist.

Along with the bulk of the Eighth District staff, I was up in St. Louis for about two months. By then, our offices in the federal building in New Orleans were functional again. I grabbed a ride south with one of the other guys on staff and moved back into our house. I had work to do.

THIRTEEN

Back in New Orleans (2005-2008)

Our house was intact, but other houses nearby had major damage from uprooted trees that landed on them or from roofs that had been blown away. Three short blocks from our house, the elevation was a little lower, and the houses had all flooded. Our house had been closed up for two long, hot months, and still did not have the electricity or gas restored. Most immediately, the refrigerator had been closed up all that time and was so bad that I didn't dare open it. A few other Coasties helped me duct tape it closed and drag it to the curb for FEMA to eventually haul away. The black liquid that dripped out of it as we dragged it down the driveway made us all gag.

The whole of New Orleans was still very much a disaster zone. Only a few grocery stores were open, and selection was pretty limited, as were the stores' hours. Efforts to clean up were getting started. A few houses had already been gutted and had piles of crap outside, but most still just looked abandoned, and there were a lot of dead cars around. These were cars that had been flooded and stranded. They were consequently left covered in a light brown muddy coating from the flood water. The entire city had a sickly sweet smell that remained for weeks until all of the dead refrigerators had been picked up and disposed of. That would eventually happen one by one, and street by street.

I got the gas turned back on at our house when I spied a worker from the gas company working the next block over as I returned home from work a few days after I had gotten back to New Orleans. He was happy to turn it back on and relight the pilot light on the hot water heater. I didn't have electricity, but I could take a hot shower and the gas stove was back in business. It took ten days after my return to get electricity restored. The power company insisted that I had to get a certified electrician to tell them that everything was fine and that our house had not been flooded. The fact that the houses on either side of mine had electricity didn't make a difference to them. Finally, another Coast Guard captain and a friend of mine who was still working in the city's multi-agency command center asked the power company representative there to cut through the red tape and turn my electricity back on. It was back the following day. I think the fact that he was a Louisiana native helped make that happen. You gotta know someone to get things done in Louisiana.

By November, the District staff were mostly back to normal working hours. We still all had our primary responsibilities that had understandably been temporarily and entirely set aside while we dealt with Katrina and her aftermath. We were all pretty busy catching up. The weekends were generally open, but they quickly filled up. I was only a few blocks from my kids' high school, Ben Franklin High, and the entire first floor had been flooded. Volunteers were working to clean it out, so I pitched in. The school administration set a goal to have the school reopened in January. That seemed ambitious to me, especially since the staff and teachers were scattered all over Louisiana and neighboring states as were the students. But they needed help, so I pitched in. On the first day, I helped the music teacher/band director clean out the band room and instrument storage areas. He got pretty emotional when he found the cellos that had been soaked in the flood water and destroyed. It had taken him several years of budget saving to purchase them. I left him in tears and quietly dragged them out to the garbage pile since they were well beyond salvage. I helped at the school for quite a few weekends and was content to just be whatever manual labor they needed.

But it reached a point where I had to step up and say something to the principal and football coach, two amazing and hard-working people who were trying so hard to execute their vision. I laid out a critical path for them to get the school open again, related to the building. Tossing out destroyed books and school records was easy, but they had to focus more on the school's electrical system, plumbing, and heating/cooling ventilation systems. I spoke with them about the critical path and sequencing of the work, and I hoped that my nudge would help. I resisted the urge to get more involved because there were just not enough hours in a day, and I did not have the local New Orleans connections that they would need to get the work done. In the end, they were able to reopen in January. They only had a portion of the student body and a portion of the teachers, but they got it going again. If only the City of New Orleans had had as much vision and drive.

We decided that Peggy and the kids would stay in Savannah for the remainder of the school year, and therefore, Conor's high school graduation from the Benedictine Military School (known locally as "BC"). There wasn't enough infrastructure in New Orleans up and running yet, and they didn't need to be there to watch the city's ineptness and unbelievably slow progress in coming back from the storm's damage. They returned to New Orleans when the school year was over, and Haley returned to Ben Franklin High the following year and graduated from there. Conor graduated from BC and went off to college in Ohio.

While I was helping at the high school, one of the other officers on the Eighth District staff put together a list of Coast Guard families whose homes needed to be gutted. As it should, the list included active duty, civilian employees, and retired Coast Guard people who still lived in the area. Family is family. So, when I wasn't helping at the high school, I pitched in to help out other Coasties. It was a lengthy list, but we started working on it one weekend day at a time. A couple of the active-duty families had just moved into their homes when they had to evacuate. Gutting their houses included hauling out moving boxes that they hadn't even had a chance to unpack before the flood waters rose. It was a sad thing to see.

Gutting houses was nasty, smelly, dirty work. Mostly nothing was salvageable. All we could do was carry or drag everything to the curb to be eventually hauled away by FEMA contractors. Furniture, appliances, clothing, kitchen stuff, books, files, everything was ruined. Most of the dressers were so swollen and warped from the flood waters that you couldn't open the drawers to even lighten the load a little. Family photo albums were particularly hard to trash, but between the mold and the water damage, nothing could be done to save any parts of them.

I helped one family who were friends from Haley's Irish dancing. They were native New Orleanians and the salt of the earth. Their two-story house was in an area that was flooded so severely that the flood water stain was two feet up the walls on their second floor. I showed up to assist them and their teenage son (the Irish dancer) with some gutting. It was especially tough because they were taking the time to look at every destroyed item and mourn the loss of the most cherished of them. When the couple ran out to get more industrial-strength garbage bags and grab some lunch for us, I kicked into overdrive and worked like a maniac to remove as much stuff from the house as possible while they were gone. Their son looked at me a bit oddly, but I think he understood and was not nearly as sentimental about the items as his parents were.

Every weekend was like that for several months until we exhausted the list. I got pretty good at stripping off nasty, smelly clothes outside the back door so I would not drag the mold spores and smell into the house any more than necessary. My dad sent me a hat that I proudly wore that proclaimed "illigitimi non carborundum" or "don't let the bastards get you down." It was something he quoted every now and then, and it applied perfectly, especially when we were actively gutting a house and a tour bus full of disaster gawkers drove by. That didn't sit well with me, but I resisted the urge to flip them off or moon them, or something else equally inappropriate. The hat was great for me, and it survived until we were done with all the house gutting and school cleanout. At that point, it was too dirty and foul to save and was ceremoniously tossed with the last of my house-gutting clothes.

Those weekends were exhausting, but I was exercising my Coast Guard option. It was the right thing to do. I had promised one reserve master chief that I would help him and a group of other Coasties on a particular Saturday with his house. Around mid-morning, I heard him wonder aloud if anyone had seen Captain Close. I started laughing because, at that moment, I was atop a pile that had been his detached garage and was in the process of dismantling the framing and roofing. He didn't know I was there because I had just shown up first thing and started working. There was no need for fanfare, formality, or recognition. There was just work to do. Plus, he didn't recognize me in my very dirty clothes and strange ballcap, and sunglasses. It was my Coast Guard option.

In the immediate aftermath of Hurricane Katrina, a lot of Coast Guard families with small kids had extra challenges. Schools were not open, and there were questions about whether or not some would reopen. Daycare providers had not returned to New Orleans, and many daycare centers were either destroyed, had not reopened, or had extremely limited staff. To their credit, the Coast Guard offered transfers to members in difficult situations because of the lack of infrastructure and support systems, but initially, the Headquarters assignment officers were reluctant to shake things up too much. It was outside the norm for them, and they were uncertain about issuing Permanent Change of Duty orders outside of the normal rotations.

While we were still up in St. Louis, I was approached by a young Petty Officer whose wife and two small kids had followed us up to St. Louis. He didn't work for me, but I was there and available to help the Coast Guard do the right thing. One of his kids had serious special needs, and they had already confirmed that the specialists and care providers would not be back in New Orleans for months, if ever. He asked for assistance getting a transfer. At my request, he provided me with a list of places with Coast Guard units around the country where the appropriate services were available. It was a short but reasonable list. I made a few phone calls to the assignment officers and was able to get him orders to one of his requested locations. It was the right thing to do, and even the small Coast Guard was big enough to handle a few extra sets of orders and transfers.

One of the biggest privileges I had in the post-Hurricane Katrina time was being part of the Awards Board at the District Office. Rear Admiral Duncan had gotten agreement from up the chain that all awards related to Hurricane Katrina efforts would go through the Eighth District's Awards Board. That would assure some consistency in awards for Coasties who came and took part from all over the Coast Guard. The only downside was that we had a lot of deserved award recommendations to review.

There has never been true consistency in formal awards and medals despite everyone's best efforts. Outstanding work can get overlooked because a supervisor never gets around to drafting the award and supporting information, or because it gets denied as it works its way through the hierarchy. Efforts related to Katrina would be no different, but largely for different reasons. The pace was so fast for so many days that actions blurred together. It was difficult for exhausted pilots to remember which rescue swimmer did specific things, and it was challenging for more senior aviators to identify which pilots were involved in specific rescues. Similarly, the boat forces involved in rescues in flooded parts of the city were often working independently and focused on the people who needed help. During the rescues, none of the awards stuff was important, and rightly so.

It was only when everyone was back at their unit and rested that they realized how hard it would be to document what took place and the truly heroic efforts of so many under very trying circumstances. They tried their best to document what happened, and I was honored to review every submission and award recommendation in detail. Few were denied, but some were downgraded. A few were upgraded to be consistent with other awards already approved. I kept hard copies of most of the award submissions because it just seemed disrespectful to throw them out.

We had only been in New Orleans for a year when Katrina struck, and by the time the next transfer season was starting, many of us were offered the option to transfer early because of the hardships associated with the hurricane's aftermath. I discussed it with Peggy and opted to stay. It didn't feel right bailing out on the unit and bailing out of New Orleans. It would have been the perfect excuse to go somewhere else and start a new tour, but I couldn't do it. Plus, I had developed a good working relationship with the inland towing industry. They didn't all like me, but we had a solid mutual respect, and I liked working with them. They deserved better than starting over with a new "Chief, Western Rivers" so soon. Plus, they had been previously treated as an afterthought on occasion and deserved better in that regard as well. Except for hurricane responses, they had my full attention all the time, as did the units up the rivers that reported to me. They deserved some continuity, so I stayed.

We continued to work closely with the inland towing industry and continued to improve our working relationship over the next two years. We worked through several high-water periods and several low-water periods and put our new Waterways Action Plan to the test. Along the way, the six Marine Safety Offices and three Coast Guard Groups morphed into three Sector commands. We tried hard to make the transition as seamless as possible and without any negative impact on the varied missions. The biggest change was that the Marine Safety Offices in Pittsburgh, Huntington, and Paducah became subunits of Sector Ohio Valley in Louisville, and their respective Captains of the Port/Officers in Charge, Marine Inspection lost a little clout in the process. A lot of the details of this change were logistical and administrative stuff, such as identifying the chain of command for writing fitness reports for all of the offi-cers. Those details ate up some time, but we didn't miss a beat operationally. One of the benefits was that I had fewer commands working for me, which made things a bit simpler.

Living in New Orleans was going to be a challenge after Katrina. It was unbelievably frustrating to see how dysfunctional the city government was. There were lots of ideas proposed related to the best way to rebuild the city. Some were a bit wacky, but most had some merit. It seemed to me that the city needed to pick one or two of the better ideas and run with them. Instead, they picked none of them and just let things play out however they would play out.

The most successful New Orleans stories involved people who just picked themselves up and made things happen. The small restaurants that reopened with very limited menus, very limited hours, and very limited staff were inspirational. I walked into one not long after returning to New Orleans. It was a take-out place, and as I was looking up at the menu, the woman said, "Honey, we don't have any of that." I asked her what she had, and she said she had two things: red beans and rice, and rice and red beans. That sounded fine to me. It was delicious.

I stopped in a local family-run hardware store in the early days after returning to New Orleans. I didn't really need much, but I admired the fact that they were even open. I was in uniform and got to talking with the owner while I bought a wheelbarrow and an axe that I correctly thought would come in useful for some house gutting work. The store was on slightly higher ground and had not flooded. He said they were a bit of an oasis from the flood waters. He said he used his Marine Corps Landing Safety Officer skills and was helping to land helos in the back parking lot. The helos were loading people who made their way to that location from flooded areas and were then flown to shelters. Semper Fi! The front windows of the store were still boarded up, and to prevent the urban search and rescue teams from breaking in to search for people or bodies, he had spray-painted on the plywood "Semper Fi," "No dead bodies," and "USMC." His store is where the city's rebuilding effort should have started, or at least the inspiration for the rebuilding effort.

Things got most of the way back to normal, but not nearly all of the way back to normal. To begin with, "normal" in New Orleans is a little different from normal everywhere else. Stores, shops, and restaurants reopened when they could. Each well-known one, like Brocato's Café with the best cannoli ever made, was celebrated by the citizenry with a line that went around the block. But some places never came back, and some people never came back. However, the drug gangs came back, and they brought with them better contacts for suppliers that they had developed in other cities while they were evacuated from the hurricane. And since the makeup of the neighborhoods changed because of the hurricane's destruction, a whole new series of turf wars developed. That crap just made a hard comeback for New Orleans even harder.

By the time 2008 rolled around, we were ready to leave the Big Easy. Conor was two years into college, and Haley was graduating from high school and would be heading off to college. I felt I had fulfilled my obligations to the inland towing industry and to the units that worked for me by providing some real continuity. I had also seen how things could work at that level in the Coast Guard; the coordination opportunities, the capability to provide real support to the field units, and the "big picture" concerns of the District Commander.

I desperately wanted to get a field command at a Sector on the coast. Several were opening that year, and I felt truly blessed to be assigned to Sector St. Petersburg in Florida. I was going back to St. Pete, where I started my Coast Guard career, and we would only be 45 minutes from Peggy's family in Sarasota. It would be the closest we had ever lived to family.

FOURTEEN

Sector Command! (2008-2010)

For the second time, I relieved a classmate of mine. This time it was Joe Servidio. Joe was the first Sector Commander there and had overseen the transition from a Coast Guard Marine Safety Office and a Coast Guard Group command into one Sector command. It was a big task and was done well, although the unit was geographically split between the primary facility and command center in St. Petersburg and the legacy Marine Safety functions that needed to be near the port in Tampa. They worked out of the old Marine Safety Office in Tampa. As Sector Commander, I had an office in each location but primarily worked out of St. Petersburg.

From day one, I loved the job. It was stressful and fast-paced, and multiple missions were always happening at the same time. The staff was good, and the crews were professional. More importantly, the leadership at the small boat stations, on the buoy tenders, and on the patrol boats was top-notch. We would all be tested.

I took full advantage of the several days of overlap I had with Joe Servidio and probably asked so many questions that I was a pain in the ass. While playing Q&A, I also got to see how operations were done and how decisions were made.

The Seventh District Commander from Miami came up to preside at the Change of Command. He was new to his job, too, and planned to stick around for a day after the ceremony to learn more about the unit and the area.

As I pulled up in my car for the ceremony in my Full Dress Whites uniform, the young petty officer asked if I was there for the Change of Command. My wife leaned over and laughingly informed him that I "was the Change in Command." I thought it was hilarious, but the young kid was embarrassed and no doubt hoped that I would not remember who he was later. As was common, the ceremony was formal and included too many speeches. The formal aspect is very traditional and time-honored, but we all know that the real change of command is when the outgoing person hands the official cell phone to the incoming person.

The next morning, I was escorting the District Commander, Rear Admiral Steve Branham, around the base, including a few buildings that should have been on the historic register, such as the warehouse that was originally a Coast Guard float plane hangar with a wide ramp right down into the water. Then I got a phone call from the Commanding Officer of the USCGC VISE, an inland construction tender. While dismantling an old aid to navigation structure that morning, one of the crew members had been injured and was being rushed to the hospital. I was impressed that the admiral and his command master chief, who was also there, did not get too excited or press for formal investigations. Instead, they appropriately let me deal with it. While the incident was serious and could have been much worse (picture a chainsaw flying through the air), appropriate safety precautions had been taken. The cause had more to do with the unknown nature of the old wooden structure, the condition of the various timbers, and how it had been constructed 50 years earlier. Fortunately, the injury was not too severe, and the young crew member recovered.

Two days later, I got a call from my command center in the evening while I was at home. Station Fort Myers Beach to my south was in hot pursuit of a boat suspected of being involved in human trafficking from Cuba. The Commanding Officer of the station was requesting permission for his boat crew to shoot the engine on the boat. The chase had begun shortly before when the suspect boat refused to stop for a boarding. It fit all the details of being part of an outbound smuggling operation, either carrying fuel to resupply other smuggling boats or carrying people being smuggled in from Cuba. The chase was happening in the dark, in mostly calm seas at speeds around forty knots.

I only really had two questions for the station's Commanding Officer: How good was his boat Coxswain, and how good was his shooter? He answered that both were the best he had. That was all I needed to know. However, the decision to shoot out the engine(s) required approval from the District Commander, and the command center was quickly in touch with the District in Miami with my strong recommendation. While waiting for their reply, I chatted with the C.O. at the station. I had only met him once before, but was impressed. When permission was finally granted, the crew fired a series of copper-jacketed slugs from a shotgun that quickly brought the boat to a stop. There were no injuries, just arrests. While waiting to hear the results of the action, I couldn't help leaning my head into the next room and quietly whispering to my wife, "We're shooting at some bad guys. I love this job."

My Deputy Sector Commander was Commander Aylwyn Young. Ayl brought great experience with Coast Guard small boat operations and administration, and just had an amazing depth of knowledge and understanding. It only took a few days on the job to realize that his experience and judgment were being overlooked. I would get calls directly from the command center and was being asked to make decisions about various search and rescue cases. I did not want to add layers to the decision-making process, but the two most experienced officers, Ayl and Lieutenant Commander Tim Haws, my Chief of Response, were not being brought in on the decisions. Also, after one night, when I got seven or eight calls from the command center informing me of each step of each search pattern, it was time to make some changes.

First, I told the command center not to call me unless Commander Young and Lieutenant Commander Haws were on the line. I wanted to hear what they had to say, and I wanted them to hear my thoughts, questions, and decisions. Believe me, it was not a group decision, but I valued their experience and recommendations and wanted that to be part of my decision. Secondly, I reduced the number of calls. I did not need to hear that our small boat or aircraft was completing search pattern A and starting search pattern B. I also did not need to make the decision to request a helicopter from the District in Miami. That was a routine request and could be made at a much lower level. Plus, I trusted the command center officers and crew. They were good, especially with search and rescue (SAR) cases, and they knew when to get excited and when to treat cases as non-critical.

My instructions were fairly simple. They were to call me when they needed a decision, especially decisions to stop searching for someone missing. If the crews found whoever we were looking for, the command center could tell me at the morning brief. If the case was still ongoing, they could update me at the morning brief and throughout the day. After announcing that I wanted Ayl in on those calls, his reaction was "Thank you." To me, it just seemed dumb for me not to include him. I valued his input in every case.

One other revelation I had related to SAR cases was that the command center watchstanders would occasionally want to guess about where the missing person was or was not based on their experience. I valued their experience, but didn't like guessing. I felt that their job was not to speculate what most likely happened, to the detriment of what could have happened. Their job was to find whoever was missing and eliminate areas by searching them, starting with the most likely areas and moving to the less likely areas. Then I would decide when enough was enough. That approach probably resulted in some extra aircraft and small boat search patterns, but that's what I would want if they were searching for someone I knew.

As I met several of the Coast Guard reservists assigned to Sector St. Petersburg, I was both impressed and perplexed. Impressed because they were professionals and rightly proud of their Coast Guard service. Perplexed because even though the Coast Guard had integrated reservists into the active duty units years before, there was no consistent plan for the reservists to gain specific skills to truly augment the unit. For example, one reserve junior officer was randomly assigned to the Prevention Department in Tampa, where he drilled a total of two days per month and two weeks per year. He was very interested in earning a qualification as an investigator for commercial maritime casualties: injuries, fatalities, collisions, groundings, etc. The problem was that active duty Coasties gained that qualification only after they had qualifications as a Coast Guard marine inspector. It took active duty people several years to gain all of the qualifications and experience necessary to become qualified investigators. At the rate of two days per month and two weeks per year, it would take this young man thirty years to develop usable skills and qualifications. I had a better idea, and it did not involve asking each reservist what he or she would like to do.

Reservists were an augmentation force. But from my perspective, we didn't need them to augment our normal day-to-day duties. We needed them to augment us during big events and big incidents like major oil spills, large security events (like the G8 Summit), large-scale exercises, and large-scale surge operations (like military outloads). The commonality for all of those big incidents was that the Coast Guard unit would morph into an Incident Command System (ICS) structure. That was something the Coast Guard has invested massive amounts of training, expertise, and effort into developing. More significantly, it worked. It was a very flexible structure with a unified command. It enabled the Coast Guard and any other federal, state, and local agencies to coordinate and work together to address incidents that required massive coordination and were often very dynamic.

Three things really made the ICS system work. First was the specific ICS training, usually done with other agencies sitting right next to

Coast Guard people and taking the same training. Second, ICS required continuous identification of daily goals, daily planning, tracking of resources, and daily tasks, all of which were assessed and fed into the development of the next day's goals. Third were the clear lines of responsibility that allowed the people doing the work to focus on that and not worry about logistics (someone else's responsibility), finding resources (someone else's responsibility), tracking the costs (someone else's responsibility), etc. Top all of that with a unified command consisting of the Coast Guard as the lead federal agency, a state agency usually, and the responsible party (the company that spilled the oil, for example). In any big incident during which we would need our reserve augmentation, we would most likely be standing up our ICS structure. My concept was to have the reservists focus the reserve officers on training for ICS qualifications we would need, and to focus the enlisted reservists on the surge operation skills we would need.

I thought it was a sound approach, and the command staff and the senior reserve officers at the unit were convinced. The District staff in Miami heard me out and then punted to Coast Guard Headquarters, where all policies for reserves were developed. The staff at Coast Guard Headquarters was non-committal. They hinted that some sweeping policy change might be coming, but had no timeframe and no real idea what the changes would be. It was pretty evident that I would get nowhere with them. So, I exercised the Coast Guard option. The reservists assigned to the unit worked for me. So, I would train them and push them to develop the skill sets that I knew would be useful and that I would need in the event of a big incident. The Reserve Program Office at Coast Guard head-quarters could catch up to me at their own pace, although I was pretty sure they were not capable of making any changes at any time during my tour at Sector St. Petersburg. That turned out to be a correct assumption.

Having spoken with the senior reserve officers at the unit, I next approached the reserve chiefs. Chief petty officers, senior chief petty officers, and master chief petty officers are senior enlisted men and women with years of experience. They are properly regarded as seasoned leaders and technical experts in their respective rates. The reserve chiefs met as a group regularly, and I wrangled an invitation. I had asked previously what the reserve chiefs had been doing during their reserve drills and found that many were being treated as technicians and not as leaders and experts. That was partly because some had little expertise in day-to-day sector activities and partly because they, too, had been asked what they would like to do. My approach was to ask them if they would like to do work that was more appropriate for their ranks and specialties. I wanted them to be leaders and responsible for the junior enlisted reservists. I asked them to get onboard with my approach so that the next time a big incident happened at Sector St. Petersburg or elsewhere in the Coast Guard, they and their people would have useful skills. They would be more than augmentation bodies that needed to be trained.

I tasked the reserve officers with identifying, planning, scheduling, and coordinating the training needed for themselves and all the other reservists. For the officers, the training would be heavily focused on ICS qualifications. For the enlisted reservists, the training would be commensurate with their specific rates but would also focus on security skills, boat crew skills, logistics skills, and the sort of skills I saw we had needed from our reservists post 9-11 and during the G8 Summit.

The Coast Guard never bought into my approach, at least not while I was on active duty, but I didn't care. What I did care about was that when we stood up a major security operation in early 2009 when Tampa hosted the Super Bowl, and when we responded to the Deepwater Horizon massive oil spill in 2010, my reservists had been trained in the ICS structure we used and had skills that were in demand. I loved the crap out of them for it and relied heavily on them with great results.

When I first learned that Tampa would be hosting the Super Bowl, I did not expect it to involve us too much. The stadium was inland and not on the water, and we were never involved in the security for the regular season Tampa Bay Buccaneers games. But what I quickly learned was that the Super Bowl is almost a full week of activities involving all of the team owners, celebrities, sports figures, and large numbers of fans. Every hotel would be full, and numerous activities would take place in downtown Tampa on the water. Plus, the Super Bowl was a magnet for the rich and famous, many of whom would bring (or send for) their massive yachts. And there would be a football game, too.

Altogether, this would be a big security operation and involve numerous federal, state, and local agencies. The planning began months in advance. Early in the process, we met a retired policeman who worked directly for the NFL as their security liaison in the Tampa Bay area. As I learned, the NFL had one such position in each city with an NFL team. The Tampa Police Department would have the lead in many ways. But the local FBI was heavily involved, especially related to intelligence information, because the Super Bowl also attracted a wide variety of criminals from across the country and would always be a potential terrorist target.

My staff quickly began planning what we would need to secure the waterways and the shoreside structures on the water. We requested additional Coast Guard resources, including parts of several specialized Coast Guard security units that came complete with their boats, weapons, and command structure. But, most of the people and equipment for the operation were from Sector St. Petersburg, including our reservists.

It was pretty close to an "all hands on deck" evolution since there was no way to reduce our normal day-to-day workload of search and rescue cases, aids to navigation work, commercial vessel inspections, etc. All of that would still be happening, plus a major security operation.

Super Bowl Sunday was 01 February 2009 that year, and there were fortunately no significant security incidents leading up to or on that day. Along with several other Coast Guard officers from my staff, I was in a temporary multi-agency command center in Tampa from well before the game started until well after. As evidence that normal Coast Guard missions were still happening, we had to conduct a medical evacuation, or medevac, of a woman from a cruise ship that had just departed Tampa just a few hours before the game started. The unfortunate woman apparently slipped and fell down a staircase and broke her leg very shortly after the cruise ship left the dock. She never even made it out of Tampa Bay before her cruise was over, and she was on a Coast Guard small boat and on her way to the hospital.

I would be remiss if I didn't point out that the Super Bowl was quite a good one that year. My Pittsburgh Steelers defeated the Arizona Cardinals 27-23. As the Steelers were driving late and down by three points, everyone in the multi-agency command center but me was cheering for the Cardinals to stop them. They did not want the Steelers to kick a field goal, tie the game, and possibly send the game into overtime, which would have lengthened the security operation. To my great joy, the Steelers did not kick a field goal but instead scored a touchdown on an amazing pass from Ben Roethlisberger and an equally amazing catch by Santonio Holmes to go up by four points and guarantee there would be no tie and no overtime. It was a nice ending to a very busy few days.

FIFTEEN

The Big SAR Case (2009)

Less than a month later, I got a call at about 0130 on a Sunday morning. We had another SAR case. The command center relayed the key information: four men in a 22-foot boat who went out fishing Saturday morning were overdue. They were well overdue, and the wife of the boat owner (and one of the four missing men) was very concerned. She said they were never this overdue, and her husband always called as soon as the boat got back into port. She had no idea where they went fishing, other than it was offshore. At some point in the evening, she had contacted a friend of her husband who frequently went fishing with him and who was an experienced boater, but he did not know where they had gone. The command center also spoke with the friend, and he indicated that the GPS on the boat had numerous locations identified well offshore where they had success fishing on other trips, but he did not have that information stored anywhere else. And the command center passed the information that two of the four men were current NFL players. While that fact was not important relative to the SAR effort, we all knew that the case would attract media attention that other cases under similar circumstances would not.

I could tell immediately that the weather had deteriorated because I could hear the wind blowing hard, but the command center added the details. A sunny, breezy Saturday with the air temperature in the sixties and one to two feet waves had changed that evening to strong winds, fully overcast skies, temperatures in the forties, and ten to twelve feet waves offshore. We were aware from the start that the sea conditions and the weather made this case a fight against the clock.

The command center had already requested a helicopter from Air Station Clearwater, which had been approved immediately. A second helicopter would be quickly launched as well. They had also ordered Station Sand Key in Clearwater to launch a small boat. Station Sand Key was the closest small boat station to where the missing boaters launched from, but it was also the right choice because they had a 47-foot Motor Lifeboat.

The 47' was a very capable, rough weather boat that was able to right itself if turned upside down by a wave or swell. It also had a fully enclosed cabin with four suspension-mounted seats with four-point harness seat belts. I had no idea why that capable of a boat was even located in Florida, as most were located along the wild coasts of Oregon and Washington and other areas where massive waves regularly formed due to the sand bars at harbor entrances and river mouths. The Coast Guard would later move that boat to a more northern location, but on that night, it was the right boat for the job, and no other small boat we had would have been safe to send out. The Commanding Officer of Station Sand Key, a CWO (Boatswain), was a solid professional and leader. He picked the four-person crew and made sure that two of them were qualified coxswains, i.e., both fully qualified and capable of driving the boat and being in charge while underway. Coxswain qualifications were not handed out but were earned.

The immediate issue was that we did not know where they had gone fishing after they left a marina in Clearwater. We did know that the boat was constructed in such a way that it would not sink. The boaters could have had engine issues on the way out or on the way back in. The boat could have flipped in this bad weather. They could have gone overboard and been separated from the boat. Having some idea of where they were headed would have narrowed the search area. But we did not have that information. And the weather was still bad.

The 47' from Station Sand Key got underway and did a straight-line search in case the missing boaters had headed due west from Clearwater. It took the 47' eight hours to travel 50 miles out and back. Search conditions for the crew on the 47' were difficult. They were bounced continuously in their seats as the boat pounded through the waves. When the boat was in a trough, all the crew could see was a wall of black water. When they were on the crest of a wave, they were limited to what their spotlight could illuminate while the crew looked intently for any other lights that might be out there on other boats. Their search was unsuccessful, and when they got back, they were all bruised from the straps that held them in their seats, so violent was the thrashing they endured.

At the same time, the two HH-60 Jayhawk helicopters were also searching in grid patterns. They used their infrared surface search camera as well as night vision goggles. They were bounced around and buffeted by the high winds. Occasionally, they took off their night vision goggles and turned on their downward shining spot-light, not so much as to search, but in the hope that anyone out there would see the bright light and know that we were searching for them. That sometimes caused missing boaters to then send up a flare or turn on a strobe light. Search conditions were difficult for the aircrews as well, with heavily overcast skies, high winds, and large waves. We were uncertain whether we were looking for a disabled boat that was drifting, an overturned boat with a much smaller profile, or persons in the water. So, they searched for all of the above.

The four missing men were Marquis Cooper, who owned the boat and was one of the NFL players, Corey Smith, the other NFL player, Nick Schuyler, a personal trainer and acquaintance of Marquis Cooper, and Will Bleakley, Nick's close friend and former teammate on the University of South Florida football team.

I went into the office very early that Sunday morning and met with the command center team. They explained the search patterns that they had used, were currently using, and were planning for follow-on searches. My only guidance to them was to do their best to ensure that we were searching continuously without time gaps.

There was solid science behind these search patterns. They were developed from a probabilistic computer model with specific details inputted by the command center watchstanders. The program could forecast where a drifting object would travel from a starting point based on the wind, waves, currents, and tides, including how they varied during the time period in question. The model also took into account the nature of the object being searched for. For example, a disabled and drifting boat would be more influenced by the wind and would therefore drift in a different direction than a person in the water, who would be more influenced by the waves and currents. The model could also determine a likely starting point based on where a floating object was found. If a drifting boat was located but no one was onboard, the model could be used to determine which direction the boat was drifting, backtrack from there, and then calculate the likely drift of persons in the water.

This probabilistic computer model would quickly calculate and produce a concentration plot based on a thousand model runs. It was a brilliant piece of software, especially in the hands of a skilled user. The art that accompanied this science came from handling the unknown factors since we had no known starting point for a drift, had not found anything yet, and were unsure whether we were looking for a boat or persons in the water. The command center SAR controllers were experts in running a series of models with varying constraints and options. They ran the model as if we were looking for an upright boat, then ran it again for an overturned boat, and then ran it again for persons in the water with and without lifejackets. They took the composite of the model run concentration plots and devised search patterns for the helicopters and fixed-wing C-130s from Air Station Clearwater, as well as for surface assets. The patterns took into account the aircraft and boat time limitations, capabilities, and weather because, in bad weather, the spacing of the legs of the grid searches had to be tighter than the spacing of the legs in good weather. On that day, with continued overcast skies, there was no need to alter the patterns to avoid having crews running lengthy search legs looking into the sun. Skies remained gray and overcast all day.

Early Sunday morning, we also dispatched the Coast Guard Cutter CROCODILE, one of our 87-foot patrol boats home based in St. Petersburg. The Officer-in-Charge was Senior Chief Boatswain's Mate Smith, a tall, thin, no-bullshit leader who ran a good boat. Initial reports from CROCODILE were that search conditions were difficult. Even the 87-foot patrol boat could only search when on the crest of the swells and waves. Nevertheless, they were given a search pattern from the command center, and they got to it.

By early afternoon, I was on the phone to the District in Miami requesting a larger cutter be sent to us that could handle the weather more effectively than the patrol boats and small boats that we had at our disposal. The District identified and redirected the Coast Guard Cutter TORNADO that was already underway and patrolling south of the Florida Keys. I alerted the command center that TORNADO was on its way, with the ETA unknown at that point.

We made contact with the families of the missing men to keep them updated. I spoke with Nick Schuyler's mother and separately with his father. Lieutenant Commander Tim Haws kept in touch with Marquis Cooper's wife and also with Corey Smith's family in Virginia. Will Bleakley's parents drove down to the base from about an hour north of St. Petersburg, where they lived. We met with them and explained what we knew and what we did not know. We went into detail about how we were searching. They remained on base for much of the day, but they were long and undoubtedly painful hours for them with no news.

I had the first press briefing on Sunday. It was mostly attended by local TV crews, and we held it on base in a small conference room.

But within a few hours, the media presence grew, and national media outlets arrived. At that point, we made all the media wait outside the gate, just off base. I did not need them wandering around the base interviewing random Coasties, and we wanted to give Will Bleakley's parents a buffer so they would not be bombarded with questions. All subsequent press briefings would happen outside the gate.

At a meeting with my senior staff to discuss the operation, Tim Haws had a brilliant suggestion. He suggested we reach out to the NFL security liaison in Tampa we worked with for the Super Bowl. We could brief him on what we were doing, then he could brief the NFL League office. The hope was that we would convey that we were doing all we could and thereby forestall any public statements or efforts by the NFL or the two teams (Marquis Cooper was with the Oakland Raiders, and Corey Smith was with the Detroit Lions) to influence our actions. Tim reached out to our guy, who was willing to be read into the operation and liaise with the NFL. We even allowed him into the command center, a rare occurrence for someone who was not a Coastie or a law enforcement official, but I felt it was warranted to let him see that we were being forthright and transparent with him.

The Coast Guard Public Affairs Detachment in Clearwater was supposed to consist of two public affairs specialists, one a first class and the other a third class petty officer. However, the more senior billet was empty, leaving only a junior petty officer to assist. But Petty Officer Kneen was excellent. She dealt directly with the media and kept them informed as appropriate. She advised me on when we should have press briefings and reviewed talking points with me. She also prepared written press releases for my approval. Despite a very youthful appearance, she was a true professional and herded the media like a seasoned pro. And there was a lot of media. Even ESPN showed up. The press briefings were just that, short briefings where I listed our ongoing efforts and resources actively engaged in the search, and then took a few questions. I tried hard to keep my answers short and to the point. Several reporters would ask a question that had already been asked, but I think they wanted to capture the answer while I was looking at their camera. That was easy enough. There were also some stupid questions ("Are there sharks out there?") and a few questions that I did not want to answer ("How long can they survive in that water temperature?"). So, I just didn't answer them and asked for the next question.

In the late afternoon, Tim Haws and I were in the command center in contact with the CROCODILE by radio. I asked Senior Chief Smith how they were doing. "Fine, sir," was his reply. Then, Tim Haws motioned to me and rephrased my question in a much better way. He asked Senior Chief how effective they were. The Senior Chief replied, "Not very." He indicated that the seas were rough, and it was difficult to search effectively. He said that all of the crew was seasick except for him. That was enough. Since they were approaching the end of their latest search pattern, I instructed him to come back in. The 87-foot patrol boats were capable platforms but were not good riding boats. If their effectiveness was limited in daylight, it would be even less once the sun set. If they were executing effective searches, they would stay out there and keep going. Instead, it would remain an air search until the bigger Coast Guard Cutter arrived or the seas calmed a bit.

I dragged myself home late that evening to find that my wife had been fielding phone calls about the SAR effort from family members who had heard or seen me on TV. It had been a stressful day for us both, and the search continued unabated. I had only been asleep for a short while when Nick Schuyler's mother called me looking for an update. I reached out to the command center for the latest actions and called her back with an update. I felt for her and understood her stress level and difficulty sleeping. It was a tough conversation as we had nothing positive to report.

I was back on base before dawn on Monday. The first thing I noticed was that CROCODILE was not tied up at the pier as expected, so I immediately went to the command center. The CROCODILE was safely anchored in the lee of Egmont Key at the entrance to Tampa Bay. Senior Chief Smith had decided to anchor in the calm lee of the island so his crew could get some rest and they could be quickly back into the search on Monday. The weather was expected to be better on Monday, and he wanted to get back into the action as quickly as possible and forego the lengthy transit from there to the base. It was a smart move, using great judgment and an inherent sense of taking the Coast Guard option.

Corey Smith's family was due in early Monday morning as well, having driven all night from Virginia. We met them in our conference room, and the cooks had food and coffee waiting for them. I never found out who had arranged for the food, but it was a heartfelt gesture and was appreciated by the Smith family. I briefed them on everything that had transpired up to that point and answered all of their questions. They were good people in a very tough circumstance.

The 179-foot Coast Guard Cutter TORNADO finally arrived after dawn on Monday morning after fighting through rough seas all night. The CROCODILE was ready to get back in the action as well. The seas had calmed somewhat, the wind had lessened to a solid breeze, and the sun was starting to come out. The search continued. Over time, we had expanded the search area partly because the possible locations of a drifting object would diverge with time and partly because, having searched areas with no results, we tried other, albeit less likely, areas. Still, the issue was that we had no starting point or even a suspected starting point for where the boat was when it became incapacitated. The three things we were certain of were that the boat would not sink, there was still hope of finding them alive, and it was still a race against the clock. So, the search continued with both surface and air assets.

The command center put the TORNADO and CROCODILE into search patterns early that morning. Two helicopters were also still searching. Mid-morning, we got a message in the command center that TORNADO had found the boat and one person. That message was sent by open VHF broadcast, which was being monitored by the media and some of the public. We instantly started getting calls asking for more information. We switched all comms to a secure channel. The TORNADO had spotted the overturned boat with one person sitting on the hull holding onto the lower end unit of the motor. The TORNADO launched their small boat with a swimmer onboard. They rescued the survivor and brought him back to the cutter. He told them his name was Nick Schuyler. He had a plastic bag with a few wallets and cell phones that confirmed who he was. The crew of TORNADO got him into some dry clothes, and their advanced first aid-trained crewmember did a quick medical assessment. Nick needed to get to a hospital, so we directed the nearest helicopter from its search pattern to medevac him from the TORNADO and take him to the hospital.

The Commanding Officer and crew of TORNADO did solid work finding him. I learned later that the boat and survivor were first spotted by the Chief Machinery Technician, who had come up to the bridge from the engine room to lend a hand. He spied "a white cap that did not disappear," as white caps on waves normally do. What he saw was the white bottom of the boat as it came into view and disappeared from view between the swells. The TORNADO snapped a few pictures of the rescue that we later released to the media. Some of the pictures made the front page of newspapers all over the country.

I was in the command center a few minutes later when we got a phone call from Nick Schuyler's father. He had heard reports from the media that we had found someone and wanted to know if it was his son. I spoke briefly with him and relayed that we had found his son alive. He yelled, "Thanks! You guys are great!" then hung up. A few minutes later, my deputy, Ayl Young, and I looked at each other and realized that we hadn't passed the news to Nick Schuyler's mother, having momentarily forgotten that his parents were divorced and not really in communication. Ayl and I went back to my office to call Nick's mother. She answered the phone and immediately asked if we had found him. I told her we had, then there was just silence from the other end of the phone. There was a lot of silence. I looked at Ayl quizzically, and he said, "I bet she fainted." Sure enough, she and her daughter called me back a few minutes later. She had fainted but was okay and was sitting down now. I assured her that we found Nick. She wanted to jump into her car and drive down to our base, but I waved her off and said we were sending him to the hospital. I would call her back in a few minutes and tell her which hospital.

We also had to convey the news to the families of the other men still missing. They were happy we had found Nick and the boat but were understandably even more concerned for their loved ones. I assured them we were not done searching.

Nick was flown to Tampa General Hospital and Tim Haws sent two seasoned search and rescue experts to interview Nick to hopefully get details about the other three missing men. They were able to interview Nick once the emergency room team had done their initial assessment. What our guys reported back was harrowing and horrific. Nick was in rough shape from being bruised and battered against the boat in the rough weather but mostly from hypothermia. The doctors indicated that we found him none too soon.

Nick was able to relay that they had been roughly 70 miles offshore. When they were ready to head back in late Saturday afternoon, they had tried several times to recover the anchor without success. Then, they moved the anchor line to a cleat at the stern and tried to use the boat to pull the anchor, which had become stuck on something on the bottom. When they pulled on the anchor line with the boat, the boat squatted at the stern, swamped, and flipped in a matter of seconds. That was Saturday late in the afternoon. They recovered three lifejackets from inside the boat, plus one seat cushion that floated, but the radio and flares were soaked and unusable. The cell phones were useless even if they had been dry, as cell phone coverage was only good to a distance of ten to twelve miles offshore. Nick recounted how the weather had then gotten bad. The four men had struggled to remain on the hull of the overturned boat, sitting toboggan-style behind one of them, holding onto the lower end unit of the motor. The seas had repeatedly submerged them completely and knocked them off the boat hull. He also indicated that they saw a Coast

Guard helicopter searching in the distance that first night, but had no way to signal it.

Based on Nick's detailed description, it was pretty clear that hypothermia had taken the other three men one at a time.

Because Nick was in bad shape, we continued the search in the chance that his recollection under those toughest of circumstances was a bit foggy. The search continued although in a much more concentrated area now that the command center SAR controllers could adjust the computer models based on where the boat was found and where men in the water would drift based on the rough timeframes we had.

At the same time, we were honest with the families that the search could not continue indefinitely. We would search as long as there was a chance any of the men could be found alive on the surface of the water. Then, we would regroup and get ready for the next search and rescue call that would come for someone else's loved one who was missing, lost, or in trouble.

Late that evening, the District Office in Miami started making noise that they would take over the case. Coast Guard policy indicated that after three days, local units like the Sectors could turn over longer search and rescue cases to the district. I suppose that made sense if you were talking about a long-distance search that would encompass days or even weeks, but in this case, it made no sense to me. I was planning on suspending the search the following day, Tuesday, if we didn't find anything more, so it was idiotic to transfer the case to the district just in time for them to suspend it. Besides, we had been in constant communication with the families since the start and had been dealing with the media face-to-face. I was tired and the whole discussion sort of pissed me off. I wanted to see it through because we all suspected how it was going to end. When the District pushed harder, I point blank asked them how they planned to deal with the media and hold press conferences from Miami when all of the media was in St. Petersburg at my gate. They didn't have a good answer. Some of the senior staff in Miami understood, and some didn't. So, I took my Coast Guard option and told them in no uncertain terms that I was going to see this through, and if they wanted to do otherwise, they would have to take the case from me, because I was never going to "give" them the case. They backed down mostly.

Tuesday was a rough day for the families of the missing men and their close friends who had also arrived on base. I provided further updates for the families and again indicated that we could not search indefinitely. Separately, we indicated to our NFL security liaison that we planned to suspend the search at the end of the afternoon if we hadn't found anything further. It was worth it to keep him in the loop, so he could keep the NFL and the respective team owners in the loop and keep them as non-participants in the search.

I scheduled a meeting for mid-afternoon with the families. Will Bleakley's parents had not been on base Monday or Tuesday, so I called them. I asked them if they could come down to the base for a briefing. Mr. Bleakley said they could, but asked why. I explained to him that I was planning on stopping the search efforts later that afternoon and wanted to give them the courtesy of telling them face-to-face and answering any questions they might have. He thanked me for the offer and said they would come down if I needed them to come down to the base, but that they knew Will was gone. I did not need them to come down, but was struck hard by this grieving man offering to make the trip if I needed him to. I was humbled.

Mid-afternoon, I met with all of the family members and a few of their friends, who included a couple of other NFL players. We met behind closed doors with no media and just a few of us Coasties. I briefly reviewed the details of the case, the large number of search patterns, and the large number of square miles searched. I noted that we still had surface and air assets out there looking. Then, I explained that I would be suspending the search when the current search patterns were complete. It was not an unexpected announcement, but you could have heard a pin drop in the room. There were some tears. I answered all of their questions and gave them time to process everything. There were a few more questions. I invited them to stay in the room and on base as long as they needed to. I also told them that I would be holding one final press briefing late in the afternoon.

Shortly after, our NFL security liaison called Tim Haws to let us know that a group of players and family members were considering hiring civilians to continue the search. I truly appreciated his heads up, as it gave me the chance to address that well-meaning but ill-conceived plan at the press briefing. We did not need amateurs in small boats many miles out to sea in conditions that were still challenging, trying to conduct searches with no experience, no equipment, no training, and no support. It would likely result in more search and rescue cases, and we did not need that.

At the final press briefing, I provided an update on the case and informed them that we would be suspending the case shortly. I again took questions until Petty Officer Kneen told the media that was enough. Some of the questions were good, some were lame. I answered the good ones and moved on from the bad ones without comment.

The following morning, I debriefed the crew of the CROCODILE and let them in on everything that had happened that they were unaware of. The TORNADO came into port as well, and I met with them. I congratulated and thanked them for their efforts. I also explained the rest of the story to them as well including all that was happening behind the scenes that they had not been aware of. I offered to do the same to the crew of Coast Guard Air Station Clearwater. Their Commanding Officer, Captain Todd Sokalzuk (now Rear Admiral, retired), and I had a great relationship. He indicated that several of his crews were upset because they had not found the boat and men that first night or on Sunday. I offered to discuss the case with his whole crew if he thought it would help paint the broader picture and a few days later I did speak with the unit. It was a tough case and even the best crews with the best technology could not find the boat and men in those horrific weather conditions.

Several months later, my Command Master Chief, Louis Rapaport, and I were planning to drive up to visit Station Yankeetown. It wasn't far from Will Bleakley's parents. We stopped in to say hello to them. I wanted to give them a copy of the briefing book that we had put together for each of the families. It contained pictures of the various assets involved in the search and copies of each of the numerous search patterns we conducted. They were obviously still grieving, but so very gracious. Informally, I shared some of the details with them of the search and the people involved. I explained how the small boat crew from Station Sand Key had gone out and gotten beaten up, and how CROCODILE had searched hard, then anchored in the lee of Egmont Key so they could get back into a new search pattern faster when the weather calmed a bit. I explained how competitive the crews are in a case like this and how each unit, whether aviation, small boat, or patrol boat, wanted to be the unit that found them. I explained how each SAR controller in the command center wants to design the search pattern that proves successful. I shared how this competitive spirit makes everyone look harder, search smarter, and handle getting pounded by the seas longer. I wasn't explaining how good the Coast Guard is, just that we did everything we could as well as we could, and that this was never just one more routine SAR case. Search and rescue is such an integral part of our Coast Guard ethos that no case is routine.

A month or so later, Will's father sent two specially made Tervis tumblers with Will's initials, his football uniform number from when he played at the University of South Florida with Nick Schuyler, and the logo from the University. One was for me, and one was for the CROCODILE. I still have mine sitting on a shelf in my office. It is a pointed, if touching, reminder that not every case is successful and that each case directly impacts people and families. This would not be the last SAR case that ended sadly for some families, but we succeeded more than we didn't.

SIXTEEN

The Kidnapped Kid (2009)

The command center got a call from the local 9-1-1 operator late one night toward the end of November 2009. A local couple in St. Petersburg who owned a marina had been watching the eleven o'clock news and saw a story about a three-year-old boy who had been kidnapped on the east coast of Florida. The boy's estranged father took him during a supervised visit. When the supervisor took a quick bathroom break, the father grabbed the boy and took off. The news story included pictures of the man and the boy, and the marina owners recognized them as having departed that afternoon on a sailboat. They called 9-1-1, who passed the call to us. Early the next morning, I got together with Ayl Young and Tim Haws. We had no clue where the sailboat went or where it might be headed. The marina operators did not think the father/kidnapper knew how to sail well. There were lots of unknowns, and it was a big ocean out there, but it just seemed wrong not to take aggressive action. So, we exercised our Coast Guard option. We took a "not on my watch" approach.

Under the assumption that the guy may be using the sailboat's motor instead of the sails, we had each of our five stations send out small boats looking near shore and in the Intracoastal Waterway, an inshore waterway that runs north-south along the west coast of Florida. We also reached out to our civilian Coast Guard Auxiliary up and down the entire coast and asked them to do the same thing. The auxiliarists have no law enforcement authority, but significantly for this case, they are located throughout the area, including places where we do not have much Coast Guard presence. Plus, the auxiliarists know their local waterways, local marinas, and local spots where a sailboat trying to hide could anchor in protected waters.

We launched a lot of surface assets looking for this sailboat up and down the coast, and they looked all day with no success. The following morning, we re-evaluated our assumption that the guy would not be under sail power. What if he was sailing and headed offshore? What if he was headed to Alabama, or Louisiana, or Texas, or Mexico, or the Caribbean? Tim Haws made some back-of-the-envelope calculations based on the expected speed of a sailboat, and the command center used that to lay out a search pattern. The weather was beautiful and ideal for large area searches with sun and flat seas. We easily got permission from the District in Miami to launch a C-130 out of Clearwater. Using onboard electronic systems, the C-130 crew could search in 25-mile-wide swaths since the weather was so good. The grid pattern covered a lot of territory. Several hours in, there was still no joy.

Late that afternoon, Master Chief Lathrop from the command center came by to tell me that the C-130 was on its final leg of the search pattern and had found nothing. It seemed like we would need to re-evaluate our plan yet again. Five minutes later, the Master Chief popped back in and said, "Captain, belay my last. They found them at the very end of the last leg of the last search pattern." The aircraft relayed information and imagery about the sailboat's size, sail shape, and color. Tim Haws used it to confirm that it was the sailboat we were looking for. The sailboat was well southwest of St. Petersburg, heading in a southern direction. Now, how do we get them?

Once again, our patrol boat, CROCODILE, was on tap. They were already underway and just had to be redirected. We coordinated with Sector Key West to our south, and they sent one of their patrol boats, KODIAK. We put a plan together quickly. It was already late in the day, so I wanted to surprise this guy at first light and catch him unaware. We arranged for both patrol boats to rendezvous with the sailboat just before dawn and kept a C-130 in the sky to watch him from a long distance and track any course changes.

We had been in contact with the local police department where the kidnapping occurred because it was their case, too. They suggested that we have someone meet the kid who knows him, and said the mother could be available. Having the mother onscene sounded like a plan fraught with emotion and uncertainty. Instead, we invited the female police detective who had been working the case and who had met the boy several times previously. She turned out to be a real trooper. Because the plan came together fast and would be executed soon, she drove well into the night to Coast Guard Air Station Clearwater. We would fly her out in a helicopter to meet one of the patrol boats once they safely had the boy. At some point, the FBI wanted to get involved and wanted to have a hostage negotiator as part of the boarding team. I refused. Our boarding teams were experienced and knew what they were doing. I did not want an unknown FBI agent acting like he was in charge and making adjustments to our boarding procedures on the fly. I compromised and offered him a seat on our helicopter. He would be in the vicinity and available should he be needed.

The crews from the patrol boats were great. As the guy came up from below deck shortly after dawn, he was greeted by two armed boarding teams with shotguns as they were about to board the sailboat. He surrendered peacefully. The boy was transferred to the CROCODILE, and the father was transferred in cuffs to the KODIAK, which also took the sailboat under tow. They headed back to Key West, while the CROCODILE headed to Coast Guard Station Fort Myers Beach. The police detective was lowered in a basket to the deck of CROCODILE, took over watching the young boy, and rode CROCODILE back to Fort Myers Beach. The boy's mother was waiting on the dock along with a news team who filmed the teary reunion. As luck would have it, the reunion on the dock happened right in front of the bright Coast Guard stripes on the bow of CROCODILE. We could not have scripted it any better than that.

Months later, when the CROCODILE had its change of command ceremony, the young boy and his mother were honored guests. The young boy had apparently been made an honorary crew member. It was great to see that mini-reunion.

SEVENTEEN

110′ Issues (2010)

At Sector St. Petersburg, I also had two 110-foot Island Class patrol boats assigned. These cutters were capable vessels but had some issues. Built in Louisiana, these replaced some older patrol boats. They were a little bigger and heavier and had better endurance. By 2008, these 110s were working hard. The Coast Guard had recently tried to extend the hull length of some of the 110s by thirteen feet but the project went sideways. The longer hulls on the first few modified patrol boats quickly developed structural issues that the shipyard could not satisfactorily fix. Consequently, the patrol boats with the lengthened hulls had to be laid up. The crews from those laid up patrol boats were assigned to the remaining 110-foot patrol boats, creating "blue and gold" crews with two Commanding Officers and two of every other crew member. The "blue" crew would go on patrol, return, do maintenance, and then hand the cutter over to the "gold" crew. While the "gold crew was on patrol, the "blue" crew got training, took leave, ordered spares, and did all the necessary administrative work.

The concept worked well enough because the maintenance periods employed both crews. Otherwise, the patrol boats were working almost twice as hard as they had previously been. Eventually, there would likely be fatigue issues with the hulls and machinery.

While the two 110s and all their crews reported to me administratively, their patrol schedule was controlled by the Seventh District in Miami. They went where the District told them to go and when they were told to go. I never had one of the 110s available for Sector St. Petersburg's use locally.

One day, we were notified that one of our 110s had developed a crack in the main deck. The crack was problematic enough that the patrol was being cut short, and the boat was returning to St. Petersburg. Any hull crack is a problem. A crack in the main deck, not due to any trauma, is indicative of a stress fracture. That's not good. Oddly, the crack was reported to run fore and aft. Normally, a stress fracture would run from side to side ("athwartships") as the hull is subject to more bending forces in the longitudinal (fore and aft) direction.

I knew I could leave the repairs to the Coast Guard Naval Engineers. They had gotten much better than when I was one of them on STEADFAST so many years before. But what fun would that be? I suspected they would send out a repair crew who would simply weld the crack closed without any investigation as to why it cracked. I had been a marine inspector and had seen and ordered lots of hull repairs, so I exercised the Coast Guard option. We would approach the problem jointly from a commercial marine inspection perspective and a Naval Engineering perspective. I didn't give the Naval Engineers any option. The boat was at my dock and was assigned to my unit.

When the patrol boat arrived, I met it with Lieutenant Commander Kevin Carroll (now a retired Captain). He was my Chief of Inspections and an expert at commercial vessel inspections. As advertised, it was a fore and aft crack directly over one of the crew berthing areas and ran two to three feet long on the main deck just aft of the superstructure. It was just big enough that rain and seawater would run down into the berthing area.

The crew also indicated that it had gotten longer since first discovered.

A close inspection showed that this part of the main deck had been previously repaired with a long, narrow insert. The current crack was almost exactly three inches away from and parallel to the previous repair. The problem was obvious to a marine inspector. The crack was in the heat-affected zone of the previous repair. In other words, the previous repair was done poorly, and the welder had overheated the surrounding metal to a temperature high enough to weaken it. As the hull, including the main deck, was stressed by the swells and waves, it wasn't strong enough to withstand normal stresses and therefore cracked. What was also obvious to a marine inspector was that no long, narrow insert would have ever been allowed on a commercial vessel. Inserts needed to be large enough that they spanned at least two frames, didn't create more stress points with ninety-degree corners, and didn't damage the surrounding metal by overheating it.

I was happy to let the Naval Engineers make the repairs, but Kevin Carroll and I laid out the extent of the repair. A large portion of the main deck would be cut out and replaced, and stresses would be minimized. All of the previous repair work would be removed in the process. Further, the welders would have to use an approved welding procedure and be certified to use the procedure. That is what would guarantee the repair would not create new heat-affected zones that would later become problems. The repair crew came down from the Coast Guard Yard in Baltimore, and they were professional, prepared, and certified.

The repair was a pain in the butt for them because they had to remove so much that was in the berthing area overhead: wiring runs, lights, ductwork, ceiling insulation, mounts for the beds and lockers, etc. It took them a few weeks to get it all done, but they did a good job of it, and the problem never returned.

Overall, it was a minor problem. But I hoped that this would be the start of the Coast Guard recognizing that they had substantial in-house expertise related to steel and hull repair and should use some of that expertise when Coast Guard cutters go to drydock. My idea was not new, but the Naval Engineering community and the marine inspection community had become pretty "stove-piped" over the recent years. I had not forgotten that the two programs had previously rotated officers between them routinely. It was what I had tried to do back in 1990 without success. As far as I could tell, sharing the expertise had ended way back then. I was happy to do my part to remind the Naval Engineers that the Coast Guard had another resource they could use if only they wanted to.

EIGHTEEN

Deepwater Horizon (2010)

On 20 April 2010, the "Deepwater Horizon" oil platform experienced a catastrophic oil well blowout that led to an explosion, fire, and an uncontrolled oil spill. Eleven men were killed and many others were injured. Crude oil, under pressure from the earth's layers deep below the bottom of the Gulf of Mexico, spewed from the well in a steady and massive quantity.

The initial impact on Sector St. Petersburg was only that the father of one of our Coast Guard petty officers worked on the Deepwater Horizon platform. It was a couple of tension-filled days for her before she learned that he was injured but safe.

The impact quickly changed. As it became clear that the volume of oil spilling was massive, we better understood the extent of the area that was affected by the spill and would likely be affected by the spill. Coast Guard Sector New Orleans responded to the oil spill quickly after the initial search and rescue aspect was over. They would need help sooner or later.

Even with a lot of unknowns, there were a few things that were certainties. First, we would be called to provide personnel, active duty and reservists from Sector St. Petersburg, to assist with Sector

New Orleans' response. Secondly, we would need to quickly get smart about the likelihood that any of the oil would come ashore or otherwise impact the west coast of Florida, with so many beautiful beaches and a reliance on tourism. Thirdly, we would need to be able to explain to the public and the politicians what the impact might be and what we were doing to minimize any impact. Those certainties paled in comparison to what Sector New Orleans was facing, but I knew that the fear of oil washing ashore on the west coast of Florida was palpable.

I took full advantage of the University of South Florida's College of Marine Science, which was conveniently just a block from Sector St. Petersburg. Over the next few days, my staff and I connected with a few professors of physical oceanography who were more than willing to share their knowledge and do so without seeming like they were lecturing a bunch of students. They explained how the shallow West Florida Shelf that extended out from shore 50-60 miles had currents that were different from those in the much deeper part of the Gulf of Mexico. The "Deepwater Horizon" was located in water a mile deep. The currents in that part of the Gulf of Mexico created a loop that stayed in the deep water. This "loop current" enters the Gulf from the south between the Yucatan Peninsula and the western end of Cuba. The current travels generally northward until it loops to the east and heads south, where it eventually bends to the east and exits the Gulf of Mexico through the Straits of Florida (South of the Florida Keys and north of the northern coast of Cuba). It then dumps into the Atlantic Ocean, where it merges with the Gulfstream that travels up the East Coast.

This new knowledge was significant to us on the west coast of Florida because that deep "loop current" does not cross over the underwater barrier of the West Florida Shelf. That told us that the likelihood of oil impacting the west coast of Florida was extremely small. It also meant that most of our stockpiled oil spill response equipment, such as oil boom and oil skimmers, would not be needed by us and could be sent to assist Sector New Orleans. Likewise, when we stood up our Incident Command System (ICS) structure, we could keep the staff small and free up personnel to assist Sector New Orleans.

Even though the likelihood of oil impacting us was small to nonexistent, I insisted that we stand up our ICS. It would allow us to have dedicated staff working on all the issues related to the spill and provide Sector New Orleans and then Sector Mobile, as the oil spread, with a point of contact that was fully manned and up to speed. It would also better enable us to coordinate with BP, which took responsibility for the spill and response, and to more easily coordinate with other federal, state, and local agencies in Florida that wanted or needed to be part of the coordinated effort.

We set up our ICS structure in some spaces provided by the USF College of Marine Science. Representatives from BP arrived soon after. About two days into this, the lead BP representative indicated that if we needed to relocate to larger spaces, we should just let him know, and he would arrange with a large hotel right down the street and rent their large event rooms for as long as we needed. Out of curiosity, I asked him what they would do if we needed to shift to the hotel, but a large wedding had previously booked the event space. He laughed and said the wedding party would have a wedding they would never forget, and it would be somewhere else! I laughed, but recognized that BP was willing and able to try to solve problems and overcome hurdles by simply throwing money at them. That didn't change anything for me or Sector St. Petersburg, and we never did need to relocate from the USF spaces. Later, when a few marine science professors complained to the media that it would take years of marine science research to monitor the impact of the oil spill on the ocean and marine life, BP announced new research grants for the academic community to study the long-term issues.

On a day-to-day basis, we were fielding and responding to phone calls reporting suspected oil washing up. We dutifully dispatched personnel to investigate and take samples. We found oil in small quantities here and there, but none was from the "Deepwater Horizon" crude oil spill. More importantly, we coordinated with every company and agency we could identify that had any oil boom that could be sent to the people responding to the spill.

Many of the coastal towns in Mississippi, Alabama, and the Florida panhandle started screaming that they needed massive quantities of oil boom to protect their coastline. Some were insisting that three layers of oil boom be placed along their entire shoreline. That seemed crazy to us. The boom was needed elsewhere, where the oil actually was or was heading. Oil boom is difficult and expensive to place and must be actively maintained, even in areas with a small probability of oil impact. And there was a finite quantity of it available at that point.

It became clear to me that the best thing we could do at Sector St. Petersburg was to maximize our outreach efforts and try to prevent our municipalities and residents from that kind of panicky reaction. I exercised the Coast Guard option and geared up the outreach significantly. Several senior politicians from Florida asked to be briefed regularly on what we were doing and how Florida could be impacted. We had meetings in the ICS spaces with the Governor, Charlie Crist, and then with Senator Bill Nelson. We briefed all the Congressional Representatives from the west coast of Florida or their senior office staff. We briefed the mayor of St. Petersburg and the mayor of Tampa. We also responded to requests from several county commissions to meet with them, and we attended a few town hall meetings to answer questions. There were lots of questions. We also started flying a few local reporters in C-130s from Air Station Clearwater out to the "Deepwater Horizon" spill site so they could see for themselves that it was a long way from the west coast of Florida and see for themselves that there was no oil near or even remotely near the west coast of Florida. That outreach became my focus, and we were largely successful. We never had the fear and panic on our coast that gripped so many towns in Sector Mobile's area and Sector New Orleans' area. That was our biggest contribution: doing darned little and ensuring everyone was okay with that.

My change of command was scheduled for 2 July 2010. The "Deepwater Horizon" spill would continue until September, and the clean-up would continue long after, but the activity at Sector St. Petersburg had tapered off significantly. Life goes on, and changes of command happen on schedule. And we had orders to Hawaii!

I was to be the Chief of Staff at the Fourteenth Coast Guard District in Honolulu. Several months earlier, when the Assignment Officer called me to tell me that I was being penciled in to be the Chief of Staff in Hawaii, he actually asked for my solemn promise that I would not retire in lieu of accepting those orders. He was serious. I laughed so loud I nearly hurt myself. Of course, I would go to Hawaii for two years. Of course, I could commit right then that I would not retire instead. Seemed like a dumb question to me, but I guess he needed to minimize the risk that his entire slate of assignments would get screwed up if I submitted a retirement letter without warning.

NINETEEN

Hawaii and the Fourteenth Coast Guard District (2010-2012)

But every good thing comes with a price tag. I was looking forward to a week or two off after a very busy two years at Sector St. Petersburg. Instead, we flew to Hawaii on Saturday, the day after the change of command ceremony, and I reported to my new job on Tuesday morning. My 36 hours of "rest and relaxation" were spent trying to figure out what time zone my body was in.

The interim District Commander in Hawaii was Rear Admiral (lower half) Steve Mehling. He had been the Chief of Staff there and had just been promoted to one-star admiral. He was serving as the District Commander temporarily until Rear Admiral Charlie Ray arrived. He was coming in from an assignment with U.S. Forces Iraq. So, I quickly relieved Steve Mehling of his Chief of Staff duties and got busy learning about the Fourteenth Coast Guard District that stretches east halfway to California and west to Guam, with other Coast Guard units located in Singapore, Japan, and Saipan.

We stayed in the Hale Koa Hotel for a few weeks until the house designated for the Chief of Staff was ready for us. The Hale Koa was great, right in the heart of Waikiki, and was an excellent introduction to Hawaii. When the house was ready, we moved in and then waited several more weeks for our household goods to arrive. We were roughing it until then, but "roughing it" in Honolulu is not too bad.

A quick historical note about our little Coast Guard housing complex. It was located east of downtown Honolulu, surrounded by local residential areas. The complex dated to before World War II, when it was a Coast Guard Communications Station, and allegedly was the last radio installation to be in contact with Amelia Earhart before she disappeared in 1937. In 2010, the small compound housed six senior Coast Guard Officers and their families in two duplexes and two single-family houses. As Chief of Staff, I was in a house that had been offices and an infirmary at some point, although it looked like a house. The house had no heat and no air conditioning, but it had lots of windows on each side that let in the trade winds every day. No other heat or air conditioning was needed.

Meanwhile, I was drinking from a fire hose at work, trying to meet the staff, learn about the subordinate units, the types and pace of operations, and a thousand details unique to the Fourteenth District.

For example, a month after I arrived, I represented the U.S. Coast Guard at the Pacific Island Forum. It is an annual forum of island nations and Pacific Rim countries. That year, the meeting was hosted by Vanuatu in their capital, Port Vila. I had to get up to speed quickly on the issues and concerns of various small island nations and be prepared to make commitments (or not) related to expected discussions and requests. The State Department did the overall coordination, and they stated official U.S. positions, but there was more to it than that.

Many of the small island nations' economies relied on their fisheries, especially their tuna fisheries. They either ate the tuna, sold the tuna, or sold the rights to fish for their tuna. Usually, it was some combination of all three. Of big concern to them was their ability or inability to enforce their fisheries laws. In simple terms, they were concerned they could not stop Chinese, Taiwanese, and Korean tuna fishing vessels from illegally fishing in their Exclusive Economic Zone (EEZ). That was a huge issue as their fishery was often being over-fished, and illegal fishing provided them no revenue for their natural resource. Most of the islands were too poor to afford vessels large enough to patrol and enforce their fishery exclusions, and often, their EEZ was very large and included the ocean within 200 miles of their shoreline. The large number of small islands that formed some of these island nations resulted in thousands of square miles of EEZ.

Part of our Coast Guard outreach with the full blessing of the State Department involved periodically sending large Coast Guard cutters, usually one of the 378-foot Hamilton-class cutters, on patrols that were intentionally designed to have the cutter meander among several of the island nations. We had formal agreements with several of them that allowed them to place one or two of their law enforcement personnel onboard our cutter as it patrolled through their waters. They used their law enforcement authority from our platform to enforce their fisheries. The island nations that did not have agreements in place were most anxious to get them signed and in place. One country explained that just having the signed agreement helped them by discouraging illegal fishing vessels from entering their waters because of the threat that a Coast Guard cutter might just show up. But the State Department could be very slow in signing those agreements. It seemed like common sense diplomacy to me. Signing a piece of paper costs nothing and does not commit us to sending a cutter. However, the goodwill gained for the U.S. was legitimate and meaningful. These were the issues that arose at the Pacific Island Forum. I never heard of Nauru before arriving in Hawaii. Still, there I was, standing in Vanuatu, talking with a senior government official from Nauru about fisheries enforcement and the value of a signed agreement. Two months earlier, I was worried about spilled oil washing up on St. Pete Beach, and now I was half a world away in a diplomatic setting, talking with foreign nations about fisheries enforcement.

And we also coordinated search and rescue efforts across a huge swath of the Pacific Ocean, including many of the same island nations we assisted with fisheries enforcement. We would get a report of an overdue canoe or small boat with two or three locals onboard who had been expected to arrive at one small island several days before. In the U.S., people get worried if a boater is late by a few hours. In the Pacific islands, no one worries until a few days have passed. In response to reports of overdue boats, we would make broadcasts, coordinate with other nearby nations, ask them to provide aircraft or ships to search, send one of our C-130 aircraft from Hawaii down range to search, and coordinate search patterns. These searches would cover thousands of square miles of ocean and would occasionally be successful. The missing locals were tough and resilient people who knew the ocean and lived salt water lives. If blown off course, they would catch rainwater to drink, fish for sustenance, and remain at the mercy of Mother Nature until found.

One such case involved one of the 378-foot cutters homeported in Honolulu. The Commanding Officer was my neighbor in our little Coast Guard housing complex. The cutter was down range doing fisheries enforcement when we got the call about some missing boaters. He was not far away, so we re-directed the cutter, sent a C-130, and gave them both search patterns. They found the missing men stranded on a tiny island. The cutter picked them up and took them home, where all involved received a very warm welcome and an outpouring of gratitude. It was such a great story that I exercised a bit of the Coast Guard option and had our public affairs staff send out press releases and announcements about the successful search and rescue. We even made the Commanding Officer available for an interview via sat phone. It was just too good a story not to get some positive media for the Coast Guard and some recognition for a cutter and crew that often performed their work with no recognition and no fanfare.

For the searches that were further downrange, we would often solicit assistance from the U.S. Navy, U.S. Air Force, Australian military, and the Japanese military. When they had aircraft or a ship available, they would willingly pitch in for search and rescue cases. That followed a long-standing maritime tradition of assisting anyone in distress at sea, regardless of who they were or where they were from.

The U.S. Navy had a helicopter squadron in Guam that our Coast Guard Sector Guam staff got to know and work with. If available, those Navy helicopters would fly search patterns for cases in the vicinity of Guam. They were true professionals and did great work despite not having much training in search and rescue. On a visit to Guam, the Coast Guard Sector Commander, Tom Sparks, took me by to meet the Navy Helo Squadron Commanding Officer. It was a chance to extend our thanks to the Navy pilots. Tom Sparks pointed out to me that the Helo Squadron was informally and unofficially referred to as "Coast Guard Air Station Guam."

This was an opportunity to exercise the Coast Guard option. After hearing firsthand about how they helped us, I offered to send one of our helicopter pilots and one of our flight mechanics from Honolulu to Guam for a week or so to conduct some training and share some pointers. The Navy Squadron C.O. was ecstatic. No one had ever suggested that before or thought that the brass would approve of something like that. It made a lot of sense, and I had the money to make it happen, and it was not a very expensive thing for what we were getting out of it. Plus, I knew, without even checking, that Air Station Barbers Point (just outside of Honolulu) would be happy to send a few of their people to Guam for a week. It was a done deal.

Not all of the SAR cases had happy endings. In one case, a young American guy was living among and assisting the local population on one of the remote islands that make up the Republic of the Marshall Islands. He and two others left in a small boat with a small outboard engine to travel to another small island. They never arrived. Our command center staff in Honolulu coordinated the search that lasted for days. We sent a C-130 from Honolulu, coordinated with other agencies and nations, reached out to commercial ships transiting the area, and broadcast to anyone else who might be out there in the area. We never found them or their boat.

As was usual, the command center staff kept the Marshall Islands government and the family of the missing American updated on what we were doing and ultimately on the decision to suspend the search effort. A few months later, the family of the missing American came through Hawaii on their way to the Marshall Islands and asked to meet with us. They wanted the little bit of closure that we might be able to provide them. On the District staff in Honolulu, Commander Mark Morin had been the point of contact for the family during the entire search and was the right person to give them a detailed briefing on the case. Thinking back to the big SAR case I had in St. Petersburg, I suggested that Mark put together a briefing book to give to the family. It would show the assets involved in the search, a detailed timeline of our notification and search efforts, and copies of the various search patterns. Mark met with the family. He was respectful, sympathetic, and detailed regarding our efforts, and the family was most appreciative of every-thing, including the binder. It was a tragic case with no happy ending.

Another interesting and typical case for the Fourteenth District involved a sailboat in distress in the middle of the Pacific Ocean between Hawaii and California. Two men and the young son of one of them were onboard when the sailboat hit some heavy weather and was demasted. They called for help, and our District command center coordinated the search and rescue case. The sailboat was well beyond helicopter range and a C-130 could only overfly them and drop equipment as needed. Using the Automated Mutual-Assistance Vessel Recovery (AMVER) system, the Command Center coordi-nated with a commercial container ship, HORIZON RELIANCE, that was not far away and asked them to assist in the rescue. The ship had to backtrack a bit from its route but willingly did so. With a Coast Guard C-130 flying overhead, they performed a textbook rescue in the middle of the night during a slight and temporary break in the weather to safely bring onboard all three of the sailors. It's one thing for a huge commercial ship to drift down on a small vessel in good weather and daylight to assist people in distress. But it's an entirely different beast to do it in the dark and during rough seas. They were true professionals who did great work in adverse conditions that they did not train for, and they successfully rescued the sailors with no injuries.

AMVER is a voluntary system where participating commercial ships provide position updates regularly so that they can be contacted if there is a maritime incident near them. There is no reimbursement, just a heartfelt "thank you" from the Coast Guard and those rescued. We knew that HORIZON RELIANCE had significantly delayed their schedule for the rescue. As payback, I asked our public affairs staff to make sure that press releases went out that included the name of the ship and that we ensured the media would be waiting for the ship when it arrived in Honolulu with the rescued sailors. Some free publicity was about all we could do, and this feel-good story was too good for the local media to pass up.

In March 2011, a magnitude 9.0 earthquake occurred off the coast of Fukushima, Japan. It created a tsunami that, together with the earthquake, created massive damage along the Japanese coast. Included in that was major damage to the Fukushima nuclear power plant.

The tsunami traveled across the Pacific Ocean and hit Hawaii before spreading to the entire west coast. The odd thing about tsunamis is that basic physics will tell you exactly when it will arrive, but the height of the tsunami can't be accurately predicted and is very dependent on the shape of the land (or island) impacted, the angle of impact, and the rise of the ocean floor in specific areas. In Hawaii, the warning sirens sounded, and we put our evacuation plans to use. Fortunately, the impact in Honolulu was minor, with the height of the tsunami limited to roughly two feet. On other Hawaiian Islands, the height of the tsunami was greater, and it caused some damage, but nothing like they had in Japan.

The damage to the nuclear power plant became the next big concern. The damage to the reactors was unprecedented. Radioactive cooling water spilled into the ocean. There was a loss of radiation containment. That triggered the Japanese government to create a nineteen-mile evacuation zone around the site.

Our initial Coast Guard concerns were twofold. First, we had Coast Guard personnel in Japan at Coast Guard Far East Activities (FEACT) along with their families. They were not in the evacuation zone and were over 150 miles away outside Tokyo at Yokota Air Base, but there was a lot of uncertainty about what was going to happen at the nuclear power plant, and the odds of catastrophic failure appeared high at that point.

Secondly, commercial shipping traffic heading to the U.S. from ports in Asia might be contaminated if they were in or possibly passed through the ocean outside of Fukushima, into which radioactive seawater had leaked or been dumped. Fortunately, this concern was minimized shortly thereafter as the radioactive seawater was limited to near-coastal areas off Japan.

It was the first concern that led to exercising the Coast Guard option again. Not long after the incident, the Department of Defense notified their personnel in Japan that if the situation remained uncertain or got worse related to radiation exposure, they would consider evacuating all military family members. Since the Coast Guard personnel and their families were housed at Yokota Air Base, they would fall under DOD authorization and policy for matters like this. I did not like the idea that our Coast Guard families would just be lumped in with the thousands of other DOD families that might be scrambling to get DOD transportation out of Japan to some unspecified location. My family had experienced a DOD-involved evacuation a couple of times when they evacuated from hurricanes heading to New Orleans. They were not pretty. But for the moment, the evacuation of dependents was just a contingency.

Then, the DOD decided to pull that trigger and announced that they would start evacuating dependents in the next few days. My boss, Rear Admiral Charlie Ray, was back on the mainland for a sequestered promotion board, and I was the acting District

Commander. I would have liked his advice and definitely the horse-power that comes with his rank, but I had little doubt about what he would want. I wanted to get our Coast Guard family members out of Japan and to a location of our choosing where we could take care of them without relying on the DOD.

So, I called Captain Jack Vogt (now Rear Admiral, retired) at Coast Guard Air Station Barbers Point and asked him to hear me out. Could they fly a C-130 from Barbers Point to Yokota Air Base and then directly to Guam if they put two separate crews on board? The first crew would fly to Japan with a refueling stop along the way, where the second "rested" crew would take over and fly the aircraft to Guam. At Yokota, they would pick up our Coast Guard depen-dents and evacuate them to Guam. Jack asked for some time to talk with his staff and do some math to see if it would work.

My next call was to Tom Sparks, our Sector Commander in Guam. I asked him if he could figure out how to house 30-40 Coast Guard dependents somewhere in Guam if I could get them there from Yokota. It was a loose plan, but I knew Tom well and knew he was an experienced Coast Guard Marine Safety guy who had been through his share of hurricanes and evacuations and all of the chaos that comes with them. I knew he'd figure it out even if it meant that every Coastie in Guam would have some house guests for some unspecified period of time.

Jack Vogt called me back a short while later and said they could do it. I told him to plan the details and be ready to go soon, but not to leave until I checked with the three-star admiral who was the Pacific Area Commander located in California. I also called our unit in Yokota and told them to get their families ready to be evacuated to Guam.

So, I next talked with my friend and classmate, Charley Diaz, who was the Chief of Staff to the three-star Pacific Area Commander. We both recognized that his boss would want to evaluate the polit-ical aspects of us doing this. I didn't care about that and figured that the DOD would not notice or really care about one Coast Guard aircraft and 30-40 dependents. To their credit, the Pacific Area Command quickly gave us permission to launch, but only for the aircraft to go as far as the refueling stop at Wake Island until they ran it further up the flagpole to Coast Guard Headquarters. I was very surprised when the Pacific Area heard back from Headquarters quickly, and we got a green light for the entire mission well before our C-130 landed to refuel at Wake Island.

This was the Coast Guard option at its finest. I had no idea about the details of the flight or whether they violated any flying time restrictions, and I didn't want to know. I trusted our aviation crews and our Coasties in Japan and Guam. They were doing a good thing, and they knew it. They brought all our Coast Guard dependents and a gaggle of family pets from Japan to Guam where Tom Sparks and his crew found them places to stay. You gotta love it when a plan comes together!

Several months later, and after all of the families had returned to their homes on Yokota Air Base and the crisis had been mitigated to a large extent, we invited several members of Coast Guard FEACT to come to Honolulu for an annual award ceremony that was being hosted for all of the federal agencies. It was a routine event, but I insisted that we continue to treat our outlying units like they were just down the road and not across the Pacific. When I greeted the guys from FEACT, I got a very unexpected hug from one lieutenant commander who said, "You know what this is for," in reference to evacuating his family out of harm's way. That nearly choked me up. Not taking care of our own was never an option.

As 2012 began, I knew that my time in the Coast Guard was drawing to an end. The selection board for promotion to rear admiral (lower half) had not selected me, and I was approaching mandatory retirement at thirty years of service. I felt I was a bit of a long shot to get selected for admiral. I had not attended a senior service school along the way which was usually considered a box that needed to be checked to get promoted beyond captain. I was all for further education, but relocating my family for a one-year program, and then relocating them again was not something I was willing to consider along the way. I also suspected that I had probably stepped on a few too many toes or just intimidated a few people too much along the way. That was okay. There were only a few choices I would have made differently if I had to.

In a sense, the Coast Guard exercised its Coast Guard option, and I retired with just over thirty years of active duty. We had a great retirement ceremony at Diamond Head Lighthouse. I followed my oft-provided advice for such events and thanked Peggy first instead of waiting until the end and getting all choked up.

We took some time off in Hawaii and made plans for our next adventure.

Afterword

Taking the "Coast Guard option" served me well over those thirty years on active duty and the four years at the Coast Guard Academy that preceded them. I bought into what the Coast Guard really was about, executing the missions. I also bought into the concept of using your "trained initiative," the notion that simply standing by and doing nothing was unacceptable, and the idea of doing what I felt was right regardless of the consequences. It was about fighting the good fight, as I saw it at those times. The Coast Guard allowed me the flexibility and authority to do that, and I will be forever grateful.

Just a note about my family. I intentionally did not share many details about Peggy and the kids. That's private. I love them dearly and am very proud of them. They got dragged all over the country with no say and quickly integrated into the communities and schools everywhere. It wasn't always easy or fun, but they were real troopers – a true Coast Guard family. Along the way, they kept me grounded and made me appreciate what love means.

It was a privilege every day to serve in the US Coast Guard. While a lot of the days were long, and much of the work was hard and stressful, the missions were good and meaningful. I have no regrets.

About the Publisher
TACTICAL 16

Tactical 16 Publishing is an unconventional publisher that understands the therapeutic value inherent in writing. We help veterans, first responders, and their families and friends to tell their stories using their words.

We are on a mission to capture the history of America's heroes: stories about sacrifices during chaos, humor amid tragedy, and victories learned from experiences not readily recreated — real stories from real people.

Tactical 16 has published books in leadership, business, fiction, and children's genres. We produce all types of works, from self-help to memoirs that preserve unique stories not yet told.

You don't have to be a polished author to join our ranks. If you can write with passion and be unapologetic, we want to talk. Go to Tactical16.com to contact us and to learn more.

All of Tactical 16's books are available on our online bookstore, T16Books.com. Visit it today to see more books from our selection of authors and to find a new adventure to read!

www.ingramcontent.com/pod-product-compliance
Lightning Source LLC
Chambersburg PA
CBHW052358030726
47599CB00014B/1115